making digital type look good

Bob Gordon

Designed by Graham Davis

Watson-Guptill Publications

New York

First published in the United States in 2001 by
Watson-Guptill Publications, a division of BPI Communications, Inc.
770 Broadway, New York, NY 10003
www.watsonguptill.com

Library of Congress Cataloging-in-Publication Data

Gordon, Bob
Making digital type look good / by Bob Gordon.
  p. cm.
Includes index.
ISBN 0-8230-2999-9
1. Type and type-founding–Digital techniques. 2. Graphic design
(Typography) I. Title.

Z250.7.G67 2001
686.2'2544–dc21   2001026656

Printed and bound in China

First printing, 2001

1 2 3 4 5 6 7 8 9 / 08 07 06 05 04 03 02 01

Picture and illustration credits:
Corbis Images: p. 8 Paul Seheult; Eye Ubiquitous; p. 12 Bettmann
Archives; p. 13; p. 176 Peter Turnley, Eugenie Dodd Typographics for
the use of The Fibonacci Sequence cover design and Frontier Dreams
invitation to a private view. The Department of Typography and
Graphic Communication at The University of Reading for p. 6 TR &
BL, p. 12 BL & BR, p. 13 TL & BR.

*Many thanks to those who kindly gave me their time, knowledge,
and support. These include:*
Alastair Campbell / John Cunningham of Adobe Systems Inc. / Simon
Daniels of Microsoft / Graham Davis / Eugenie Dodd / Robin Dodd /
Ben Duckett / Howard Dyke / Simon Emery / Marion Fink / Maggie
Gordon / Bill Hill of Microsoft / Melissa M Hunt of H Berthold /
Edward Jacob / Anna Leaver / Zuzana Licko of Emigre / Sue Perks /
Troy Reid of EyeWire Inc. / Tom Sawyer / Jeremy Tankard / Bob
Thomas of Bitstream / Barrie Tullett / Jeff Willis / Juergen Willrodt of
URW (Ikarus) / Petra Weitz of FontFont.

                                        Bob Gordon, 2001

This book was conceived, designed, and produced by
The Ilex Press Limited
The Barn, College Farm
1 West End, Whittlesford
Cambridge CB2 4LX
England

Sales Office:
The Old Candlemakers
West Street, Lewes
East Sussex BN7 2NZ

Art Director: Alastair Campbell
Designer: Graham Davis
Design Manager: Kevin McGeoghegan
Managing Editor: Kim Yarwood
Production Editor: Jannie Brightman
Copy-editor and Indexer: Phil Cleaver
Picture research: Shelley Gruendler

Introduction

Why do we need to know how to make digital type look good? There is an abundance of excellent fonts from which to choose, many of which we could say are designed to perfection. However, it is not the quality of typeface design that needs to be improved upon but the way in which good typefaces are used that needs care and attention. The manner in which well-cut clothes are worn may dramatically change their appearance and impact, depending on the wearer and how they are put together. We choose clothes to suit a certain occasion and similarly we would choose a typeface to suit the right task. Clothes, however elegant they may be, might call for some small adjustment to subtly craft them to individual needs and a perfect fit. So it is with type. The overall design might initially appear just right in most respects but, on closer inspection, a little fine-tuning could make it communicate more effectively.

The late 1960s effectively saw the end of letterpress printing for most commercial printing, and the move toward offset lithography. That change marked the end of a unique era of type design and typesetting craftsmanship. The design of typefaces during the 1960s was directly influenced by the physical nature and modular structure of hot metal composition. Alongside the transition from relief to planographic printing, which depended on photographically generated origination, came the development of phototypesetting. Entirely different from the physical nature of the letterpress process, which conveys an image to paper by means of a raised, inked surface, phototypesetting allowed the imaging of character shapes onto photographic material through the use of light and lenses, which opened up exciting opportunities for type manipulation.

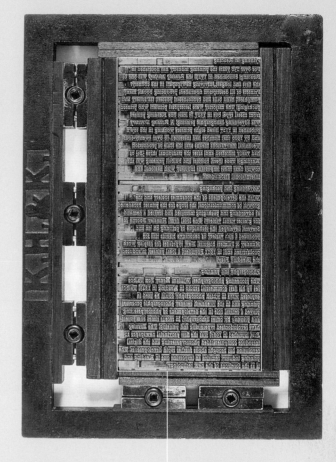

The physical nature of metal type imposed a tight constraint on page layout and inevitably resulted in rectilinear compositions.

Aluminum litho plates no thicker than a postcard, displaying a right-reading image. They drastically altered the way printers and designers prepared material for print.

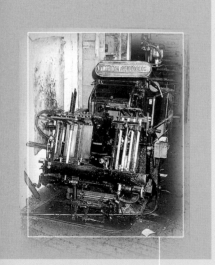

Letterpress printing machines had to be built sturdily to deal with the weight of assembled lead type contained in large steel formes.

Soon after the introduction of phototypesetting, computer-aided keyboarding and word processing emerged, often from companies previously not involved in type and print. Inevitably, this led to the unleashing into the market of innumerable pirated typefaces masquerading under well-known names, in addition to well-known types with new names. It was still possible to obtain good typesetting, but it was often expensive and difficult to find. Overall, the expansion of phototypesetting technology brought some confusion to the development of type and typesetting.

In time, phototypesetting moved on from optical methods of imaging onto photographic paper to methods using digitized fonts. Digitization considerably improved the quality and fidelity of type forms. Typesetting companies began to realize the market was hungry for good quality fonts and typesetting, and began to build libraries of interesting, well-digitized typefaces.

The arrival of the personal computer provided unprecedented opportunities for type designers, typefounders, and developers of typesetting software. Combining digital outlines, described in the form of Bézier curves, with high-resolution output enabled the creation of fonts drawn to a very high level of detail for accurate reproduction. The quality that had been thought lost had been retrieved and improved upon. Opportunities

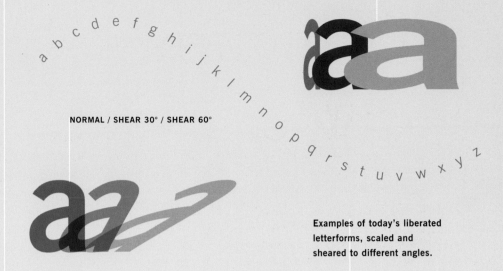

NORMAL / SHEAR 30° / SHEAR 60°

Examples of today's liberated letterforms, scaled and sheared to different angles.

for creating new and innovative typefaces were taken up with enormous enthusiasm by many designers, ushering in the start of an adventurous new era in type design and typography.

There are now virtually no reprographic constraints in getting type onto paper. We may do with type more or less what we will, and perform as many feats of typographic gymnastics as appropriate. However, with flexibility comes the sobering realization that the printed word still needs to be read. Reprographics may have gone through a fundamental change, but the need for legibility remains the same. Our range of typographic techniques now include, for example, setting text around circles and at angles, or distorting and overlapping characters, and we have many more fonts from which to choose; still we have to engage in the task of setting continuous and display text so that the message is communicated to, and understood by, a defined audience. Many of the fundamental, underlying principles of how people read, and of how we may influence this, still prevail. The only difference today is that the process can be more enjoyable.

The art of the type designer is one that embraces creativity, the rigors of craft and execution, and a thorough understanding of space, form, and the effect of delicate visual nuances. Great care, and countless hours of hard work, go into the creation of a successful typeface. Every aspect of a letterform is carefully considered, from

RIBBON 131 BOLD

CENTAUR

JOKERMAN

Typefaces have their own distinctive moods, expression, and subtle (and not-so-subtle) connotations.

Some typography has to work hard to communicate complex information in concise and comprehensible units.

E | a

km	km	km	treinnummer	717	519	521	1773	525		7	D 215		727	829	529	1777	833	
0			Amsterdam CS V	5 39	6 11	6 40	7 14			7 37	7 45	7 55	Adam C	8 13	8 33	8 40	9 14	9 32
4			Amsterdam MP	5 44	6 21	6 46	7 20			7 43			Adam M	8 19		8 47		
6			Amsterdam Amstel	5 48	6 27	6 50	7 24			7 47			Adam A	8 23	8 41	8 52	9 23	9 40
39			Utrecht CS A	6 16	6 41	7 14	7 44		8 10	8 13	8 24	Ut CS	8 44	9 01	9 15	9 45	10 01
			treinnummer	9619	6190	21	1773	623		25			627	9529	29	631	9533	D 1127
0			Den Haag SS V	5 25	6 02	6 34	7 01			7 31			HaSS	8 05	8 05	8 40	9 10	9 14	9 34
3			Voorburg	5 29	6 06	6 38	7 05			7 35			Voorb	8 09	8 13	8 44	9 14	9 18
12			Zoetermeer	5 36			7 12						Zmeer		8 20			9 25
28			Gouda A	5 47	6 22	6 54	7 23			7 51			Gouda	8 25	8 32			9 36
28			Gouda V	5 48	6 23	6 55	7 24			7 52			Gouda	8 26	8 35			9 39
44			Woerden	5 59			7 14						Woerd		8 47			9 49
60			Utrecht CS A	6 13	6 42	7 14	7 44		8 11				Ut CS	8 46	9 00	9 17	9 47	10 01	10 08
			treinnummer	919	619	21	623	9525	25	7			627	9529	29	631	9533
0			Rotterdam CS V	5 29	6 02	6 36	7 02	7 06	7 33				Rdam C	8 06	8 09	8 40	9 10	9 13
5			Rotterdam Noord	5 35	6 08			7 11					Rdam N		8 14			9 18
10			Rotterdam Alexander	5 40				7 16					Rdam A		8 19			9 23
24			Gouda A	5 53				7 26					Gouda		8 29			9 33
24			Gouda V	5 55		7 00		7 27					Gouda		8 35			9 39
40			Woerden	6 08			7 38	7 57					Woerd		8 47			9 49
56			Utrecht CS A	6 17	6 39	7 11	7 40	7 50	8 08	8 14			Ut CS	843	9 00	9 14	9 44	10 01
			treinnummer	717	519	521	1773		525	1777			727		529	1777	
39	60	56	Utrecht CS V	6 26	6 45	7 17	7 48	7 55	8 16	8 23	8 28	Ut CS	8 49	9 06	9 20	9 50	10 06	10 10
51	72	68	Driebergen-Zeist	6 35		6 54	7 26	8 05	8 25			Drieb		9 15			10 14
73	94	90	Veenendaal-de Klomp			7 14	7 39	8 22				Vdaal		9 30			10 32
80	101	97	Ede-Wageningen	6 53		7 21	7 46	8 29	8 43			Ede W	9 14	9 38	9 44	10 14	10 39
97	118	114	Arnhem A	7 06	7 33	8 00	8 25	8 42	8 54	8 57	9 05	Arnh	9 26	9 50	9 56	10 26	10 51	10 45
			treinnummer	4423			4329			4431			4333			4337	
97	118	114	Arnhem V	7 14	7 45	8 06	8 33	8 46	9 03			Arnh	9 33	9 56	10 03	10 33	10 55
116	137	133	Nijmegen A	7 30	7 58	8 22	8 49	8 58	9 16	9 21	9 38	Nijm	9 48	10 08	10 16	10 48	11 07
			treinnummer	1771			1773			1777			727				
97	118	114	Arnhem V	7 12			8 28			8 59	9 09	Arnh					10 28	10 47
128	149	145	Emmerich A	7 32			8 47			9 17	9 28	Erich					10 47	11 06
162	183	179	Wesel	8 01			9 12				10 00	Wesel					11 13	11 31
189	210	206	Oberhausen Hbf	8 22			9 32				10 17	Oberh					11 30	11 49
189	210	206	Oberhausen Hbf V	8 47			9 35				10 27	Oberh					11 42	
206	227	223	Essen Hbf	9 07			9 53				10 46	Essen					11 58	
251	272	268	Hagen Hbf A	9 52			10 38				11 25	Hagen					12 40	

the thickness of strokes to the sweep of a curve and the style of serif, from the relationship of countershapes to the juxtaposition of one character against another. These almost imperceptible differences are key influences on designers when they make their selection of one typeface in preference to another.

exercise control over the presentation of type is governed by an appreciation of, and respect for, the type's qualities and characteristics, together with a knowledge of how to control the variables in setting to achieve the desired outcome. Although we acknowledge the skills and innovation of the type designer, the use of type is a collaborative endeavor. But, unlike other collaborative endeavors, the typeface designer and typographer rarely, if ever, meet or have any interchange whatsoever over a project. The type designer offers a set of carefully designed and crafted tools to an invisible market, trusting that they will be used sympathetically, and, in turn, typographers select these tools trusting that they will perform well.

Typography is in many ways a unique art form. It may be used as self-expression, but is more usually used to communicate effectively on behalf of others. A range of graphic components used in layout, presentation, image selection, and image making is used extensively in print and other media, but none is so vital as the assembled message created by the collected icons we call type or letterforms. These give a voice to words.

Printed text is the medium by which we are able to add tone, color, character, pitch, and volume to our message, our communication, our delivery. In essence, the printed word is visible speech. Considering the style, weight, tone, and disposition of words to use on a page can be likened to preparing to deliver a message orally. How appropriate would loud or whispered words be in a

However meticulously a typeface might be refined, it would be impossible to take account of every text-setting situation. The same type design will need to be set in various sizes, to various column widths, and with various leading values. Inevitably there will be occasions when designers will wish to fine-tune word and character spacing to maximize readability while maintaining the integrity of the chosen typeface. The technology that has helped to liberate the type designer has, through numerous typesetting and page-layout applications, made available many fine-tuning techniques to permit unprecedented control over the assembly of text and display type. These controls may be used to enhance a piece of setting or, conversely, if used without due care and knowledge, may produce disastrous results. Given that the typefaces we choose are well designed in their own right, in looking at how to make digital type look good we aim to examine how, through the various controls available, we are able to create effective and cohesive typography. Unlike those of us involved in typography, most readers of text matter will appreciate good typographic design only subliminally, and they will not be aware of the hard work, care, and attention to detail that have gone into a particular piece of typographic design. This is how it should be— really good design should never impose itself on the reader. As every designer and typographer knows, it takes considerable endeavor to reach this level of typographic detailing and layout.

Making digital type look good is about understanding the way in which the printed word is to be read and about having valid reasons for making typographic modifications. The extent to which one should

Nowadays, the power of typesetting, layout, color, printing, and publishing can be at anyone's fingertips.

better weekends start from london stansted

go munich from £68 rtn venice from £75 rtn iceland from £118 rtn

book now at **www.go-fly.com**

conditions and £3 credit card fee apply · selected flights · Go uses Venice Marco Polo, Iceland Keflavik

Above: GO Airlines base their easily recognizable publicity material on a sans serif typeface, circles, and color. Like Adrian Frutiger's typeface for Charles de Gaulle Airport, and Edward Johnston's typeface for the London Underground, these types are very functional and legible.

Despite a phenomenal growth in screen-based information, there remain plenty of situations where the "printed" letterform plays a dominant role in our visual experience. Above, GO Airlines publicity material is type based.

particular situation? When should the pitch be raised—or even imitate song—with the exuberance demanded by the message? When might the tone be kept more sober? If we were to consider who might be the appropriate person to deliver a key presentation orally, qualities of voice and personal style would be of paramount importance. Should we look for someone who speaks interestingly, amusingly, or seriously with authority and, most important, clarity? Once the presenter has been chosen, we should give him or her a few guidelines, including the freedom to persuade and impress, then stand back and trust that person to do the job well. This is how we should treat type—as our presenter, our voice.

In the following chapters we look at the anatomy and terminology of type, thoroughly investigate how to choose our own preferences for typesetting values, and learn all the different ways in which those values may be controlled. Examples show the effect of the appearance of different typefaces in a mix of different column widths and various leadings.

The power of new digital technology has radically changed the working life of the designer but not necessarily made it easier. It has opened up opportunities for creativity and change in an ongoing manner unthought of a few years ago—which is exciting and stimulating. In the process it has put a greater burden of responsibility on the graphic designer, who is now accountable for virtually every aspect of his or her product. Not long ago, many processes were carried out by a host of other professionals, following instructions and adding their specialized skills to ensure the perfect job. Just in terms of attention to detail, there were practicing designers who would not care to admit, and might not even realize, that much of their typesetting had been quietly tidied up by a thoroughly professional typesetter. Now they are faced with the daunting minutiae of when and where to hyphenate, how to avoid widows and orphans, how to punctuate properly, where and when to put in a foreign accent, how to close-space words, and how to access fractions and special characters.

The paperless office has not yet arrived, but vast quantities of documents are currently held and transmitted digitally. Despite the prodigious rise in the use of mobile phones, videophones, and pagers, it seems that much information technology, now and well into the future, will remain visual and significantly word based. Where words are important, so are good layout and typographic detailing. The responsibility for making digital type look good is going to be high on the graphic communicator's list of priorities.

abcdefghijklmnop

PART 1

KK f
MMf
KERN
BASELINE

Laughter is the shortest distance between two people
Laughter is the shortest distance between two people
Laughter is the shortest distance between two people
Laughter is the shortest distance between two people
Laughter is the shortest distance between two people

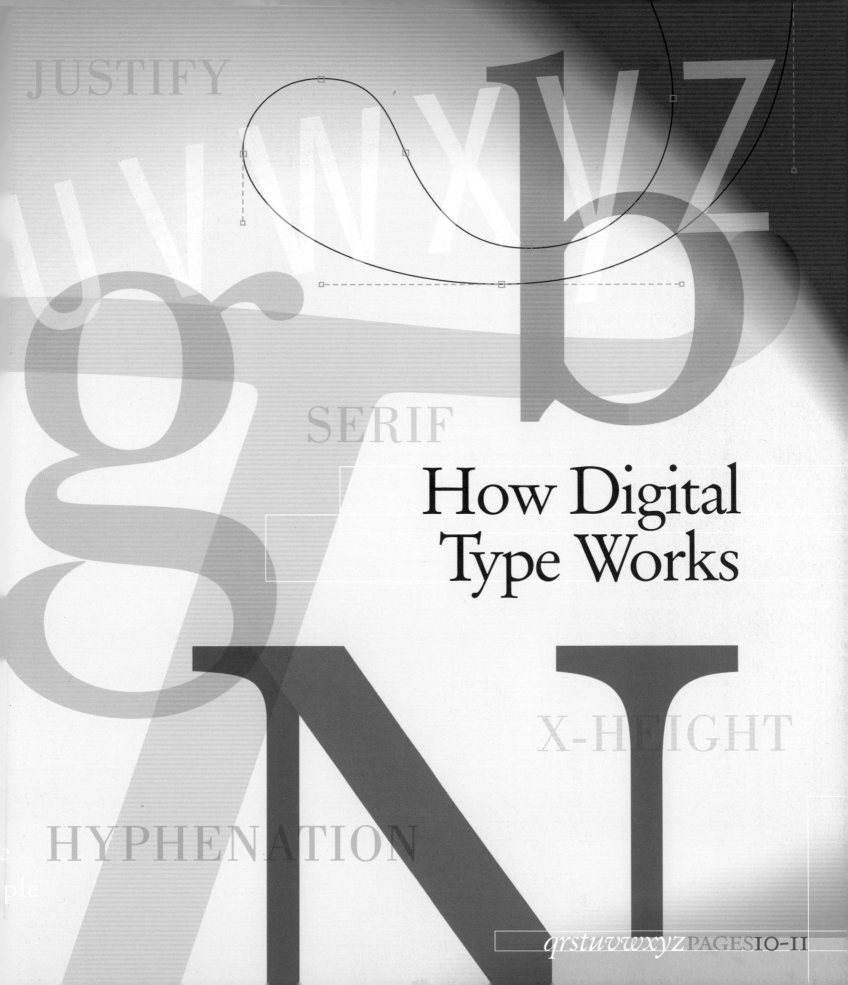

JUSTIFY

SERIF

How Digital
Type Works

X-HEIGHT

HYPHENATION

Type's rich heritage

Type's rich heritage

Efficient written language requires order and structure to produce clear understanding. This principle has been in evidence since the earliest forms of visual communication, and continues to influence our current notions of layout, spacing, alignment, and graphic presentation.

The Latin alphabet that we know and use today has its roots in Greek letterforms that were adopted and modified by the Romans. The uppercase letterforms we use today can be traced back to Roman capital letters, often carved in stone, which have endured and contributed their timeless proportions to typeface design for more than 2000 years. The roots of the lowercase letterform lie in Caroline minuscules—a uniform handwriting style developed in the reign of Charlemagne, in the early 9th century. In both upper- and lowercase letterforms, the early influence of the wide-nibbed pen can still be seen in the thick and thin gradations of upright and curved strokes. At the invention of movable metal type, attributed to Johannes Gutenberg in the 15th century, these two letterforms made the transition from handwriting to print.

The use of capitals and upper- and lowercase continues today. By its nature, metal type imposed certain disciplines on the typesetter, known as a compositor. In hand setting, pieces of type were kept in large, flat, wooden cases with compartments for each letter of the alphabet. Capital letters were held in one case, which was placed above another containing the minuscules. This gave rise to the terms "uppercase" and "lowercase." Hand setting was a laborious task in which the compositor, holding a composing stick in his hand, set lines of type by selecting each letter from the type case in front of him found on the sloping surface of a type case cabinet. The closest line spacing was achieved when each line of type was placed directly under the previous line, with the two lines touching: this was "setting solid." To space out lines of type, lead strips of various thicknesses were added between lines, from which comes the term "leading" (which originally meant added space, but in this book has its informal modern meaning of the distance between two lines of type). A skilled compositor could set around 1000 characters or letters an hour—equivalent to approximately 150 words.

At the end of the 19th century, mechanized methods of assembling type were introduced, the most notable being the Linotype machine designed by Ottmar Merganthaler. Instructions typed on a

A detail from the Rosetta Stone, showing Greek letterforms, which formed the basis of the Latin alphabet.

Example of well-incised Roman capitals at the base of the Trajan Column.

The medieval scribes paid great attention to layout, spacing, and visual interest, as well as were adept at fusing image and letterforms inventively.

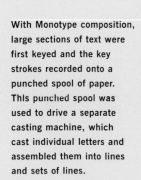

Hand composition was an exacting skill where every word and letter space had to be inserted manually.

Over the years the layout of a type case has seen modifications and variations. The early type case sat in front of the compositor as a pair on a sloped cabinet. The upper case contained all the capital letters, while the small letters were held in the lower case.

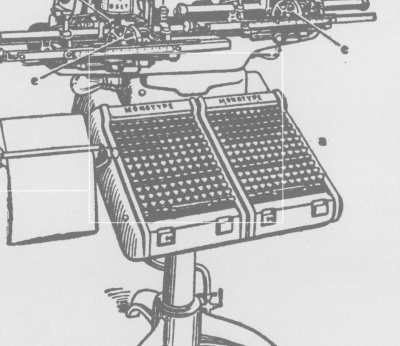

With Monotype composition, large sections of text were first keyed and the key strokes recorded onto a punched spool of paper. This punched spool was used to drive a separate casting machine, which cast individual letters and assembled them into lines and sets of lines.

With the invention of the Linotype machine, text was keyed a line at a time (hence the term "lin-o-type"). Individual matrices were assembled and cast as a single "slug" of type.

keyboard sent brass matrices of letters into a mechanical assembly system that built up a full line of text. Word spaces were wedge shaped, and when the line was nearly full the wedges were forced up to space out the words for justification. The matrices were pressed into a mold, and molten metal was then poured in to form a "slug" (a line of type), which was mechanically carried to a "galley" (an unbroken page of type). Meanwhile, matrices and spaces were distributed for use in more lines of type. A skilled operator could produce as many as six lines a minute—much faster than hand setting. Soon afterward, Tolbert Lanston was granted patents for a mechanized composition system, called the Monotype, which assembled lines and sets of lines out of single pieces of type that were controlled by a punched paper tape that was produced on a separate keyboard. These two systems (and their imitators) gave rise to the term "hot metal," which is generically used to refer to automated letterpress composition.

For many years, imprecise terms were used to describe the size of type, but in the final years of the 19th century the Anglo-American point system was universally adopted in the United States and the United Kingdom. This basic unit was 0.013837 inch, or about one-seventh of an inch. There are thus 72.27 points in an inch, although this is now simplified in most PC-based typesetting software to 1/72 inch. When applied to characters, the point size referred to the dimensions of the metal type ("body size"), which was almost always larger than the character itself and is not a consistent guide to the size of the printed letters. The same system is still used, even for types that were never cast in metal. A larger unit of 12 points, the "pica em" (often now confusingly simplified to "em"), was used to measure column width and depth.

Letterpress printing gave way to offset lithography during the 1960s. Although lithographic printing had been practiced much earlier, developments in photographic plate

With the arrival of photocomposition, designers no longer worked with hot metal types and formes, but with bromide or film, assembling the page elements on a drawing board.

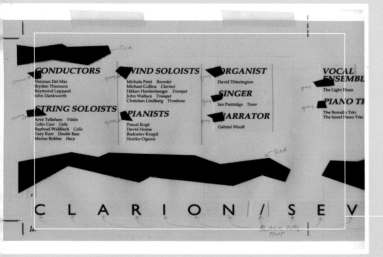

Paste-up (or artwork) was usually assembled on white boards, ready for a process camera to record the image onto film. Overlays, on translucent sheets, carried images to be printed in other inks. Further overlays carried written technical directions.

A depth gauge (or typescale) is used to calculate linefeed and for casting off (copyfitting). Working out how much a volume of typescript would make if set to various values used to be a routine chore for typographers, and the depth gauge (typescale) shown below was an indispensable aid.

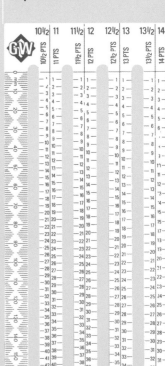

making and chemistry meant that offset lithography (the transfer of ink from plate to blanket to paper) could increase print speeds dramatically. The idea of setting type photographically had been around for many years but had never seriously rivaled hot metal. In the early years of offset lithography, type was still set using hot metal to produce "repro pulls" that were used as camera copy for the photographic platemaking process. But, as computers became more widespread and affordable, phototypesetting—the imaging of type by photographic processes—quickly took off. Text was assembled into lines by software that provided word-processing facilities such as cutting and pasting. For a couple of decades, phototypesetting developments were fast and furious, with companies coming and going, and new and enhanced equipment emerging into the market with bewildering frequency.

At its simplest, phototypesetting developments went through two distinct but converging phases. In the first phase, characters were held in negative form on disks or strips of glass or film, and were exposed onto photographic paper by means of high-speed pulses of laser light that shone through them in response to directions given electronically from a keyboard. The operator used a modest non-WYSIWYG display to view the typed text, and had a limited amount of word-processing features to call upon. Moving lenses enabled different sizes to be set from a single master drawing of each character, and photographic paper was fed vertically for each new line of type, giving rise to the term "linefeed." Text was produced in one long column of type, also called a galley, which then had to be cut up and pasted into position for the process camera and platemaking. In the second phase, typesetting software displayed keyed text on a screen, in the correct style and size of type, and assembled the text into laid-out pages. The letterforms were now generated from fonts stored as digital files, displayed on a cathode ray tube.

A lithographic plate is right reading since the image is transfered first onto the blanket cylinder and then from the blanket to the paper. In letterpress printing, the type was wrong reading since the raised ink surface was in direct contact with the paper.

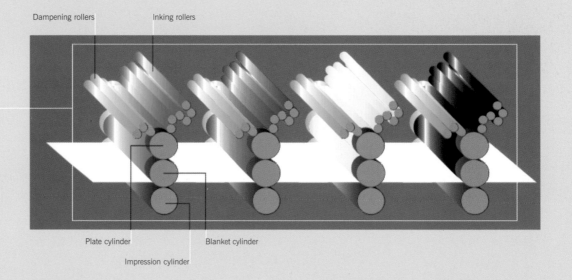

Dampening rollers   Inking rollers

The offset lithography process, showing inking and dampening rollers, plate, blanket, and impression cylinders. The four-color printing process is, in effect, four single-color presses bolted together. Many press manufacturers will add further units so that spot colors or varnishing can be undertaken on the same run.

Plate cylinder   Blanket cylinder

Impression cylinder

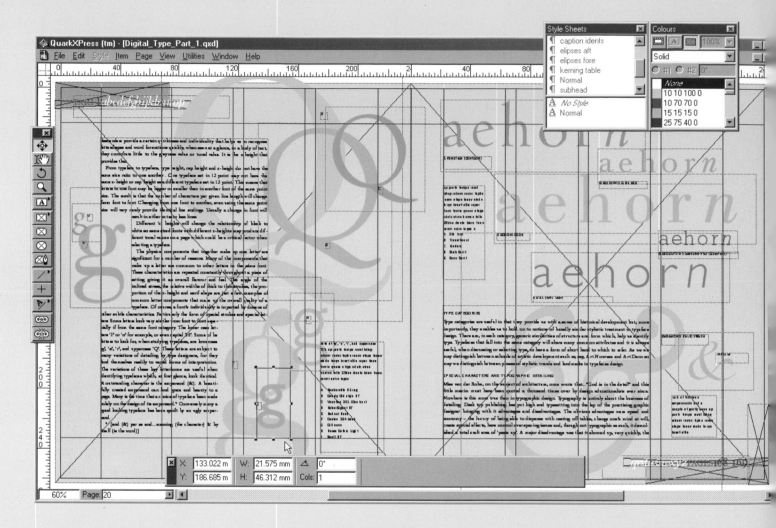

A sample spread from this book (*see* pages 20–21) displayed in QuarkXPress. The rectangles represent individual items that can be dragged around the page at will, providing a free and flexible way in which a layout can be produced. It is this freedom that distinguishes desktop publishing applications from the more structured and somewhat constraining aspects of word-processing applications. The printed spread is shown opposite.

Over most of this period much of the marketing by computer companies of phototype-setting equipment described them with undue emphasis on word-processing features, to the detriment of the quality of type output. Ease of use encouraged virtually anyone with keyboard skills to set type, and this adversely affected quality. Fortunately, designers, typographers, and publishers clamored for better-quality type and there were soon companies selling phototypesetting equipment on the basis of quality of output. It was at this time that the idea of measuring type by the metric system came up. Scangraphic marketed a high-quality typesetter that could work in a number of measurement systems, including millimeters, and (taking a lead from the German typefounder Berthold) the company's extensive type specimen books sidelined the traditional measurement of type by body size and showed it as a combination of capital-letter height (CH), minimum linefeed (LF), and the ratio CH/LF. To many this seemed eminently sensible, but it was too revolutionary for most in the business to break out of the concept of points and picas, and the point system of measurement has been retained to this day.

Once the computer was harnessed to text assembly and typesetting, it was not long before Compugraphic used an Apple Lisa personal computer to drive phototypesetters. Apple's marketing of personal computers, with a user-friendly graphic interface that technophobes could happily use, was a turning point in graphic design and publishing. Almost overnight, the origination of text that incorporated pictures could be done on one's own desktop—the birth of desktop publishing (DTP). Soon after, Microsoft launched its Windows graphic interface for PCs, designed for the office environment and with much emphasis on software for word-processing and office document processing. By contrast the Macintosh was aimed fairly and squarely at the design and publishing market. These computers gave designers almost unlimited power and control over their work.

New, exciting software companies quickly provided software for drawing, scanning, and manipulating images and, importantly, typesetting combined with page layout. Disks could be sent to bureaus with imagesetters to produce bromide prints of fully assembled pages. Sending work to specialized typesetters quickly became a thing of the past. Work produced on a computer was soon being sent down the line for immediate output to film, cutting out many traditional processes and those who worked on them.

The arrival of easily accessible typesetting was a watershed in the history of type composition as it opened up the highly professional and exacting craft of typesetting to all. Although it horrified traditionalists, it was an emancipating, democratic, and, indeed, welcome move. It encouraged many people to focus more closely on how to maintain the exacting standards of good typographic design in accepting —and being party to— innovation and change.

The Internet has created an enormous, dynamic new market for designers and publishers, but has also brought many constraints with it, necessitating innovative ways of thinking, and a push to exploit the medium fully. It is hard to believe that this explosion of digital technology has taken place in barely the last decade and a half.

Many aspects of graphic reproduction, as well as the process of origination, have gone through radical change. Where letterpress printing has given way to offset lithography, direct digital imaging onto plates is already a fast-developing technology, as is total digital printing direct to paper. Speed of processing digital information has enabled great strides to be made in many forms of origination, imaging, color fidelity, and transmission of data. Four-color process work has been joined by six-color (hexagraphic) printing, and halftone imaging by stochastic rendering—all of which have contributed to the changing world of print.

The same spread as shown on the facing page, but in finished form.

Anatomy of type

There are two fundamental aspects of the anatomy of type that dictate how we utilize it. The first is the somewhat mechanical and practical aspect of its physical dimensions, the methods by which it is constructed, and the units by which we measure it both horizontally and vertically. We need to know from which point a letter, word, or line is measured, and what the terms governing the measurement system are, so that, on a practical level, we can make our typesetting software do what we ask of it. The second is the shape, construction, and visual appearance of individual letterforms. Being able to name the type parts that make up a character's unique quality gives us a language through which to express our opinions, evaluations, and judgments.

Much of the terminology we use today comes from the days of letterpress printing. Characters were in those days constructed

Different typefaces can have different proportions within the same body size (i.e., type size).

on a physical "body" or piece of metal. The dimensions of the body included the distance from the top of the tallest ascender in the type to the bottom of the lowest descender, plus a small amount of extra space, which varied from type to type. The extra space or "body clearance" ensured that, when type was set solid (without extra line spacing), there was no danger that ascenders and descenders would touch each other, if an ascender happened to be positioned below a descender. Body size is the same as type size, our current term.

"Cap height" is the height of the capital letter. In many cases, this is not the same height as the ascenders in a given face; it is usually slightly less than the ascender height. All letters sit on a "baseline," with descenders falling below. The baseline position is of great significance because it is the point from which all digital typesetting software measures the relative vertical position of type. Line spacing (commonly now called "leading," as it is in this book) is calculated from baseline to baseline.

A key measurement in lowercase letters is the x-height, the height of a lowercase "x." It is the x-height that largely dictates the tonal value of lines of type. Paradoxically, although ascenders or descenders provide a certain quirkiness and individuality that helps

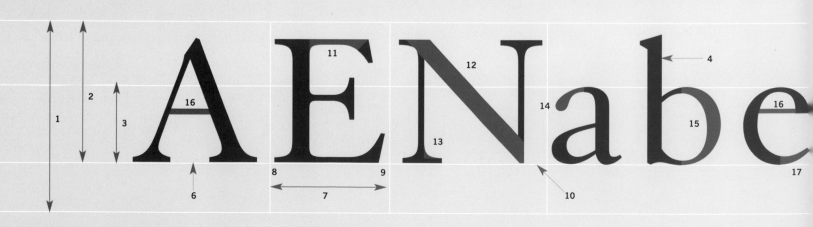

bbb

These three blocks of text are all set to the same body size and leading. The relative typographic "color" changes dramatically from face to face.

brevt elliu repar tiuve tamia queso utage udulc vires humus fallo 25deu Anetn bisre freun carmi avire ingen umque miher muner veris adest duner veris adest iteru quevi escit billo isput tatqu aliqu diams bipos itopu 50sta Isant oscul bifid mquec cumen berra etmii pyren nsomn anoct reern oncit quqar anofe ventm hipec oramo uetfu orets nitus sacer tusag teliu ipsev 75tvi Eonei elaur plica oscri eseli sipse enitu ammih mensl quidi aptat rinar uacae ierqu vagas ubesc rpore ibere perqu umbra perqu antra erorp netra 100 at mihif napat ntint riora intui urque nimus otoqu cagat rolym oecfu iunto ulosa tarac ecame suidt mande onatd stent spiri usore idpar thaec abies 125sa Imsep pretu tempu revol bileg rokam

op pretu tempu revol bileg rokam revoc tephe rosve etepe tenov sindu turqu brevt elliu repar tiuve tamia queso utage udulc vires humus fallo 25deu Anetn bisre freun carmi avire ingen umque m muner veris adest duner veris adest iteru quevi escit billo isput ta aliqu diams bipos itopu 50sta Isant oscul bifid mquec cumen ber etmii pyren nsomn anoct reern oncit quqar anofe ventm hipec or uetfu orets nitus sacer tusag teliu ipsev 75tvi Eonei elaur plica os eseli sipse enitu ammih mensl quidi aptat rinar uacae ierqu vagas ubesc rpore ibere perqu umbra perqu antra erorp netra 100 at m napat ntint riora intui urque nimus otoqu cagat rolym oecfu iunt

op pretu tempu revol bileg rokam revoc tephe rosve etepe tenov sindu turqu brevt elliu repar tiuve tamia queso utage udulc vires humus fallo 25deu Anetn bisre freun carmi avire ingen umque miher muner veris adest duner veris adest iteru quevi escit billo isput tatqu aliqu diams bipos itopu 50sta Isant oscul bifid mquec cumen berra etmii pyren nsomn anoct reern oncit quqar anofe ventm hipec oramo uetfu orets nitus sacer tusag teliu ipsev 75tvi Eonei elaur plica oscri eseli sipse enitu ammih mensl quidi aptat rinar uacae ierqu vagas ubesc rpore ibere perqu umbra perqu antra erorp netra 100 at mihif napat ntint riora intui urque nimus otoqu cagat rolym oecfu iunto ulosa tarac ecame

AMERICAN GARAMOND

CASLON 224 BOOK

SWISS 721 LIGHT (HELVETICA)

Naming parts of type lets you describe elements of letter construction accurately and thus develop a sense of structure and form.

Type parts

1 Body size
2 Cap height
3 X-height
4 Ascender
5 Descender
6 Baseline
7 Body width
8 Left sidebearing
9 Right sidebearing
10 Character origin
11 Arm
12 Stroke
13 Bracket
14 Ball
15 Bowl
16 Bar
17 Terminal
18 Finial
19 Spur
20 Serif
21 Link
22 Ear
23 Hairline
24 Counter
25 Stem
26 Spine

mgufons

us to recognize letter shapes and word formations quickly when seen at a glance in a body of text, they contribute little to the grayness value or tonal value. The x-height provides that. Different x-heights change the relationship of black ink to white page, so that same-sized fonts with different x-heights will produce different tonal values on a page. This could be a critical factor in selecting a typeface.

From typeface to typeface, the ratios of type size to cap height and x-height, and of cap height and x-height to each other, will not be constant. One typeface set in 12 point may not have the same x-height or cap height as a different typeface set in 12 point. This means that letters in one particular typeface may be bigger or smaller than in another typeface of the same size. The result is that the number of characters in any given line length will change from typeface to typeface. Switching from one type to another, even of the same point size, will rarely provide identical line endings. Often, a change in type will result in either fewer or more lines.

The physical components that together make up one letter are significant for a number of reasons. Many of the components that make up one letter are common to other letters in the same type. These components, when repeated constantly throughout a piece of setting, become characteristics, giving it an overall flavor and feel. The angle of any inclined stress, the relative widths of thick and thin strokes, the proportion of the x-height, and serif shape, are just a few examples of common letter components that make up the overall quality of a typeface. Of course, a type's individuality also depends on dozens of other subtle characteristics, particularly the forms of special strokes and special letters. Some letters look very similar in many types, especially if they are from the same type category—the lowercase letters "l" or "o" for example, or even the capital "H." Some letters to look for, when studying typefaces, are lowercase "g," "a," "r," and the capital "Q." These letters are subject to many variations of detailing by type designers, since they readily lend themselves to varied forms of interpretation. The variations of these key letterforms are useful when identifying typefaces that may, at first glance, look identical. Another outstanding character is the ampersand, "&." A beautifully crafted ampersand can lend grace and beauty to a page. Many is the time that choice of typeface has been made solely on the design of its ampersand. Conversely, many a good-looking typeface has been spoiled by an ugly or inappropriate ampersand.

TYPE CATEGORIES

Traditional type categories are useful in that they provide us with a sense of historical development but, more

Typeface recognition can often be difficult when typefaces appear superficially similar. Certain letters (characters) have a more individual form than others and it is the detailing of these that often gives you clues for easier recognition. Look at the lowercase "g," "a," "r," and "q" or the capital "Q." The design of the capital "G" is often a telltale sign in sans serif faces.

1   Baskerville BE Regular
2   Goudy Old Style BT
3   Venetian 301 (Centaur)
4   Baker Signet BT
5   Bodoni Book
6   Caslon 224 Book
7   Gill Sans
8   News Gothic Light

aehor*n*

1 VENETIAN (CENTAUR)

aehor*n*

2 BASKERVILLE BE REGULAR

aehor*n*

3 BODONI BOOK

aehor*n*

4 GEOMETRIC SLAB SERIF 703 (MEMPHIS)

aehor*n*

5 GILL SANS LIGHT

Much work has been undertaken in categorizing typefaces over the ages, with many volumes being published on this topic alone. The examples shown here offer the broadest of type groupings.

1 Old Face
2 Transitional
3 Modern
4 Slab Serif
5 Sans Serif

importantly, they enable us to develop notions of broadly similar stylistic treatment in typeface design. Each category is characterized by general similarities of structure and form that help us to identify type. Types that fall within the same category will share many common attributes, and it is always useful, when discussing or selecting type, to have a form of shorthand to which to refer. As we may distinguish among schools of artistic development such as, say, Art Nouveau or Art Deco, so too we may distinguish among phases in stylistic trends and landmarks in typeface design.

SPECIAL CHARACTERS AND TYPOGRAPHIC DETAILING

Mies van der Rohe once wrote, on the subject of architecture, that "God is in the detail." His little maxim must have been quoted thousands of times by design educators ever since. Nowhere is this more true than in typographic design. Typography is entirely about the business of detailing. That DTP has put high-end typesetting at the fingertips of the practicing graphic designer, has brought advantages and disadvantages. The obvious advantages are speed and economy: the luxury of being able to dispense with casting-off tables, to change one's mind at will, to create special effects, to have control over spacing issues, and (though not a typographic issue) to do it all without the craft skill of "paste-up." A major disadvantage is that DTP very quickly shows up the lack of accurate detailing skills possessed by many designers. Much subtle detailing used to be routinely taken care of by expert typesetters or compositors who had followed rigorous apprenticeships. Consistent, correct punctuation was supplied by the compositor, as was the correct use of ellipses, em dashes, asterisks, daggers,

GARAMOND
ITALIC SWASH

SNELL BT

The ampersand can make a typeface irresistibly beautiful— or render it unusable.

spaces after periods and sensible hyphenation. Much of this attention to detail was either not recognized or not appreciated by graphic designers. These issues surfaced when the quality of typesetting became the responsibility of the Macintosh or PC operator, designer or not.

DETAILING

Special characters that all graphic designers should be aware of are ellipses, accented letters, ligatures, fractions, nonlining numerals, small caps (small capitals), swash characters, special symbols, em dashes, en dashes, quotation marks, parentheses, brackets, and other forms of punctuation. Where special characters are involved, most digitized fonts do not have them. If the special characters do exist, e.g., ellipses, diphthongs, and ligatures, most applications require the use of special keys or commands to access them.

It is a shame that most fonts have been manufactured without at least some basic fractions such as one-half or one-quarter. One can make false fractions by tweaking characters that do exist, and some software (e.g., QuarkXPress) can do this automatically. But the results generally look thinner than the surrounding type and are not a good visual match. There are, however, Expert Set fonts that include real, well-proportioned fractions.

Expert Set fonts also provide alternative numerals, known as Old Style or nonlining numerals, parts of which fall below the baseline. Apart from being beautiful characters in their own right, they provide useful distinction when the numerals "1" and "0" are in close proximity to the capital letters "I" and "O." Most Expert Sets also contain small caps, crafted to match both the x-height and the tonal value of the typeface. Again, automated tweaking allows small capitals to be produced by scaling down normal capitals, but in the vast majority of cases they are

These two pages show just a few of the characters available from Expert Sets. They can add elegance and variety to type without disturbing the overall color and texture of the basic letterforms.

EXPERT FRACTION

EXPERT SWASH

3/4

Am

EXPERT LIGATURE

ffi

1234567890

EXPERT SMALL CAPS

AMAZON

Am

QUARK SMALL CAPS

Expert Sets have alternative characters, e.g., swash letters and nonlining numerals, and useful extra characters, e.g., fractions, ligatures, and ornaments.

EXTRA CHARACTER

too thin. Some typefaces have been designed with alternate characters, perhaps more ornate or with swashes. These are also found in Expert Sets. At present there are too few such fonts.

Set at display sizes, many special symbols such as ® and © look far too big; they look better when reduced to the size of running text and raised ("shifted") from the baseline. Some fonts (e.g., ITC Officina) come with such symbols already reduced and raised.

Poor punctuation is generally a result of inconsistent authorship and subsequent setting. A proliferation of unnecessary commas, mixed colons and semicolons, and incorrect use of single and double quotation marks, easily mar a good run of text, and are particularly annoying after you have lavished time and care over all the other aspects of esthetic detailing. Inconsistent use of en and em dashes, brackets, and parentheses can also break up the textural rhythm, thereby spoiling an otherwise clean piece of setting.

Most fonts contain properly designed quotation marks and most major DTP applications have "smart quote" features to allow them to be used easily. But all too often one sees instead the use of "tick marks"—i.e., ' and "—usually used to denote feet and inches. The type designer will have carefully and lovingly designed real quotation marks to match the style of his typeface; they should always be used. Some applications allow automatic smart quotes to be switched off. Make sure that everyone who processes your work (e.g., editors and printers) has the software configured to your specification.

Control of character and word spacing, line and paragraph spacing, will be discussed in depth in Part 2 of this book. But it would be right to make it clear here that only one word space is used after a period, and not two. "Type two spaces after a period" was drummed into generations of typists because for monospaced typewriting it improved clarity. But professional typesetters, who have always worked with proportionally spaced characters, have always used just one space after a period. Now that all computer users work with proportionally spaced letters, they should follow suit in order to avoid unsightly "holes" appearing in their text matter.

problem

HYPHEN

Fellow countrymen-Four score and seven years ago our fathers brought forth on this continent, a new nation, conceived in Liberty, and dedicated to the proposition that all men are created equal.

TIGHT SPACE

Now we are engaged in a great civil war, testing whether that nation, or any nation so conceived and so dedicated, can long endure. We are met on a great battlefield of that war. We have come to dedicate a portion of that field as a final resting-place for those who here gave their lives that that nation might live. It is altogether fit and proper that we should do this.

OPEN SPACE

DOUBLE SPACE

This example shows how a double space after a period makes an unnecessary gap in the flow of a piece of text. Justifying text across a narrow column can create over-narrow or over-wide character spacing. A hyphen has also been used in place of an en or em dash, either of which would have been more elegant.

ANATOMY OF A PARAGRAPH

Paragraphs are the essential building blocks of running text. Without paragraph breaks, text would be very difficult to read, so it follows that we should pay special attention to how paragraphs are formed. The designer should question and make decisions about the following paragraph features before any setting takes place:

- Column width
- Font
- Size
- Weight
- Style (roman, italic)
- Case (uppercase, large and small caps, or upper- and lowercase)
- Line spacing (leading)
- Character spacing (tracking)
- Paragraph spacing
- Alignment (ranged left, ranged right, centered, or justified)
- First-line indents or hanging indents
- Hanging punctuation
- Raised or dropped initial capitals
- Hyphenation

Quite a lot to think about! We cannot let decisions on these matters go by default, for all aspects of paragraph design will have a profound effect on tone, texture, page coloring, and overall effectiveness of our work.

HANGING PUNCTUATION

RAISED OR DROP INITIAL CAPS

FIRST LINE INDENTS

PARAGRAPH SPACING

FONT
WEIGHT
STYLE
CASE

SIZE

LINE SPACING

EM DASH

HYPHENATION

CHARACTER SPACING

ALIGNMENT

HANGING PUNCTUATION

"**F**ellow countrymen—Four score and seven years ago our fathers brought forth on this continent, a new nation, conceived in Liberty, and dedicated to the proposition that all men are created equal.

Now we are engaged in a great civil war, testing whether that nation, or any nation so conceived and so dedicated, can long endure. We are met on a great battlefield of that war. We have come to dedicate a portion of that field as a final resting-place for those who here gave their lives that that nation might live. It is altogether fit and proper that we should do this.

But, in a larger sense, we can not dedicate—we cannot consecrate—we can not hallow this ground. The brave men, living and dead, who struggled here, have consecrated it, far above our poor power to add or detract. The world will little note, nor long remember, what we say here, but it can never forget what they did here. It is for us, the living, rather, to be dedicated here to the unfinished work which they who fought here have thus far so nobly advanced. It is rather for us to be here dedicated to the great task remaining before us—that from these honored dead we take increased devotion to that cause for which they gave the last full measure of devotion—that we here highly resolve that these dead shall not have died in vain—that this nation, under God, shall have a new birth of freedom—and that government of the people, by the people, for the people, shall not perish from the earth."

ABRAHAM LINCOLN
Gettysburg, November 19, 1863

Full-page diagram of a
paragraph fully annotated.
Everything shown here needs
consideration by the designer.
No detail should be ignored.

Rendering type

W hether we work with type for print or for online publishing, the computer will be the medium on which we will compile our material and generate our designs and concepts. Whatever we compose on screen will be replicated by means of other processes and technology. Some knowledge of graphic screen display and graphic reproduction is needed if we are to be assured that the final output, as seen by our target audience, is exactly as we intended.

BITMAPS AND VECTORS

There are two distinct ways in which image information may be stored digitally: either in bitmap form or as vectors.

Bitmap

Whether on screen or in digitally generated print, text is rendered by means of coloring a mosaic-like grid of dots, called pixels, to form the desired shape. This arrangement in grid form is a bitmap. Bitmaps can be very memory-intensive, since files must contain information about the color and position on the page of every single pixel. If bitmap images need to be enlarged, the software can only multiply up the number of pixels, so that in effect the original pixels and all the jagged features of the original size are enlarged.

Vectors

Vectors describe a shape as an outline by means of straight and curved lines that connect strategic points. Straight lines and curves (the vectors) are recorded on file as mathematical formulae. The resulting shape is then filled with pixels. This allows for good-quality scaling. When a vector image is enlarged, the coordinates of the vector points are repositioned but the mathematical formulae controlling the straight lines and curves between those points remain constant; the new shape is then filled with pixels resulting in a smooth-edged, well-defined image. Vector image files are very much smaller than bitmap files.

Image shapes may be recorded as a bitmap or as vectors, and this applies equally to type. Drawing with vectors gives an economy of representation, since coordinates for each element are all that's needed to represent the drawn shapes. But, however it is stored, it has to be rendered on screen as a matrix of dots—in other words it has to be converted to a bitmap. Outline (vector) information must also be converted to a bitmap for printing, the difference being that the printed bitmap is of a much higher resolution.

The Bézier curve—so fundamental to the mathematical description of shapes—allows for accurate scaling for screen rendering and printing.

Vector image. Bézier curves, being the mathematical expression of a curve or line between two points, are not affected by image scaling. The distance may vary but the formulae remains constant.

A bitmap image, as the word implies, is a map or grid of pixels, the abbreviation for "picture elements." On enlargement, these pixel just get bigger and so appear cruder.

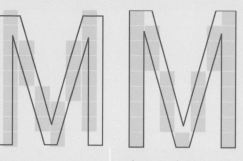

The "M" on the left has pixels arranged as close to the outline information as can be interpreted, but it looks odd. With hinting, as shown on the right, abnormalities are ironed out, resulting in a more uniform and recognizable character.

This 2x enlargement shows a true linear inch, as seen on a Macintosh monitor compared to an inch on a PC monitor. Because the PC monitor is 96 ppi, the "inch" is larger than a linear inch.

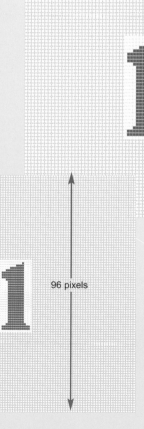

MAC SCREEN

72 pixels

PC SCREEN

96 pixels

So how does this affect the designer? Much of the designer or typographer's work is associated with making fine adjustments to detail and, for work intended to end up in print, it is not possible to rely on the screen display to see this; the screen cannot faithfully represent the correct color, resolution, character shape, and spacing. Printing, too, may not come up to expectation because laser printers and inkjet printers cannot always generate a document with exact fine detail. It is better to see printed proofs made by the same equipment that will be used for the final output.

THE COMPUTER SCREEN

An image appears on a computer screen by means of tiny colored dots, called pixels, that are arranged in a tight grid. The quality of on-screen rendering of digital type is constrained by the resolution of the monitor, which is 72 pixels per inch (ppi) on a Macintosh and 96 ppi on a PC. Small sizes of type display badly on screen, which can be especially annoying, as it is small type sizes that are most commonly used for text work.

The relatively low resolution of the computer screen is unable to display fine detail so that, when small type is rasterized and fitted to a coarse grid, details smaller than one pixel must be either enlarged or omitted. This causes such features as stroke weights and serifs to become inconsistent or be lost and results in badly formed letters and spaces. Some technique is needed to enable the rasterizing software to find ways of adjusting the arrangement of pixels to take account of lost detail. This technique is known as hinting, and hinting information is contained in all good fonts. When rasterizing software attempts to align an outline shape to a grid of pixels, only those pixels that fall within the outline shape will normally be turned on. If there are too few pixels to display a small letterform correctly, hinting instructions are used to make adjustments that will provide for a better-looking letter shape and maintain optically accurate spacing. How well hinting is carried out by differing font types may be seen by comparing them on screen. The employment of antialiasing techniques (*see page 29*) also goes a long way to improving the appearance of small type.

PRINTING

Much finer results may be achieved by printing on paper. The printing device uses tiny black or colored dots to form the image. The number of dots per inch (dpi) that an imagesetter can output is significantly greater than that of a monitor, and therefore smoother images may be formed. Most inkjet and laser

INKJET PRINTER

The quality of reproduction from inkjet printers is getting better, and in many instances can equal that of laser printers.

LASER PRINTER

Imagesetters, which image onto photographic materials, or direct-to-plate, create the most accurate images, particularly of letterforms at small sizes.

IMAGESETTER

printers are able to output at 300 to 600 dpi, and imagesetters can output at up to 3000 dpi. Inkjet and laser printers nowadays produce good results, but a close look will show the formations of dots that make up the image areas. These printers have limitations of resolution, grayscale, and color rendering. By contrast, the dots used by an imagesetter are so tiny that it is virtually impossible to detect them with the naked eye. As a result, the edges of pictures and type look razor sharp.

Page layout and typesetting applications produce files that carry sufficiently detailed specifications to take advantage of the high resolution of imagesetters, even if we cannot see this quality on our computer screens. A similar reservation applies with 300 dpi laser proofs: since higher resolution means finer output, features (especially hairlines) that look strong at 300 dpi may hardly be visible at 3000 dpi. If our work is destined for print, these devices' inability to reproduce small type accurately, because there are too few pixels with which to draw fine detail, may disconcert us; it is best to obtain printed proofs to satisfy ourselves about the quality of our work. When our work is to be viewed on 72 or 96 dpi monitors like our own, what we see on our screen is an accurate proof.

PostScript

PostScript is Adobe's patented page-description language. It converts vector information into high-resolution bitmaps for high-quality rendering. Adobe Type Manager (ATM) uses PostScript to create superior bitmaps from outlines for screen display.

Inkjet Printers

Microscopic dots of ink are sprayed onto paper at the rate of 300 to 1500 dpi. As well as being affected by limitations imposed by resolution and paper, most inexpensive inkjet printers lack PostScript, so that the rendering of type may not be truly accurate. Newer inkjets with PostScript or PostScript simulation can produce very good ouput.

Laser Printers

Most laser printers have PostScript software installed, which will successfully interpret the shape of letterforms and reproduce spacing accurately. Although they produce reasonably crisp images, lasers do have their limitations when imaging small type. Because particles of toner powder are used to create the image, smaller, lightweight letterforms tend to fill out more and appcar bolder than they actually are.

Imagesetters

A PostScript-driven imagesetter, using fast-spinning mirrors and laser beams gives the best imaging results, utilizing photographic exposure combined with very high resolution in order to replicate fine detail. Imagesetters expose light onto film, which will in turn be used to expose light onto a printing plate. The image now on the plate is inked and it is this image that is transferred onto paper.

ONLINE VIEWING

Though it is important for print designers to have a good preview of their work, the quality of screen display becomes all the more urgent for online viewing where the screen image actually is the finished product.

It is clear that a viewing device that has resolution of 72 ppi (Macintosh) or 96 ppi (PC) cannot render type as well as a printing device, although improvements are constantly being made to make small type look smoother and easier to read on screen. To draw a letter successfully, there must be enough pixels to construct the details that make a character unique. At small sizes, there are not enough pixels to do the job adequately. Only those pixels that fall within the character's shape will show. At small sizes this means that much of the letterform is discarded and much of the style and detailing is lost. To counteract this, larger type is often used for online pages than would be appropriate for a comparable printed job, but this does not really solve the problem. To address this antialiasing techniques were developed.

Antialiasing puts intermediate pixels around the edge of an image (in this case a letter), to create an impression of smoothness. This is particularly useful when used on small type.

This numeral "5" is constructed using the outline information contained in the font file.

ANTIALIASING

Sometimes, using small type is the only possible way to display more information in a given space. To allow this to happen without undue loss of legibility, software developers have created the antialiasing technique, in which the software compares the outline with the bitmap grid and works out what parts of a given letterform have become lost or distorted. The software then strategically inserts several pixels of varying gray shades around the letterforms to create an illusion of relative stroke bulk and smoothness. It is an eye-fooling exercise but it works quite well. Adobe InDesign uses antialiasing by default to display text on screen. Microsoft Windows software now smooths fonts above a certain size, depending on the font used. The downside to antialiasing is that small text looks slightly less black, but that is more than compensated for by the improvement to the integrity of the viewable letterforms.

Font technology

Font technology

In the short space of time during which digital printing and digital communication have flourished, great strides have been made in font technology. This has promoted an enormous flowering of exciting and innovative type design and has appreciably enriched the world of communication design. However, font-related issues have probably been the cause of most design-to-print headaches. A short explanation might help you to avoid such problems.

Operating systems such as Macintosh and Windows, in conjunction with typesetting software, need to be able both to display type on screen, and to send accurate type information to a printer or similar output device. To generate characters effectively on screen, the system must transform the shape of letters (glyphs) into a bitmap display of pixels to match the low resolution of the monitor. In the early days of digital type, font information was stored as fixed-size bitmaps; for each design of type there were several sizes, such as 9, 10, 12, 14, 18, and 24 point. These bitmap fonts provided the information needed by the system to display type on screen. If an intermediate size was needed, the system had to scale the bitmap information up or down from an installed size resulting in clumsy characters that were less accurate than those rendered at fixed sizes. For printing onto film or paper the letterform again needed to be converted into bitmap form but, this time, at the much higher resolution achievable by printers.

TYPE 1 FONTS

To counteract the problem of clumsy bitmapped fonts, Adobe developed fonts, called Type 1 fonts, which use Adobe's PostScript page description language. These fonts consist of two parts—a set of fixed-size bitmap font files for screen display and a PostScript

Below left: type rendered on screen using bitmap fonts of fixed (10, 12, 14, 18, and 24 point) and intermediate sizes, compared with (right) type rendered by Adobe Type Manager from vector information.

The type image above shows the rendering of various sizes of type on screen using ATM in 10% increments from 10 to 270 point.

font file, which is an outline description of character shapes. The operating system uses the bitmap fonts for drawing characters on screen, adapting an adjacent size for intermediate sizes. But, when a document is to be printed, the PostScript file containing the outline letterforms is downloaded to the printer. These outlines, constructed from vectors, are infinitely scalable and maintain perfect character shapes. Once the printer receives the scaled outline (vector) information, it can fill the shape with pixels, thus creating a new but much higher resolution bitmap for imaging. This process is called rasterization (from the German word for screen, *raster*). PostScript creates the raster image from outline information—hence the name PostScript font.

For each style of typeface—plain, plain italic, bold, bold italic—a separate PostScript file is required in addition to the fixed-size screen fonts. This makes a set of PostScript fonts appear somewhat cumbersome.

Creating Multiple Master instances in Adobe Type Manager Deluxe. Sliders in the dialog box allow for almost infinite variation across the axes.

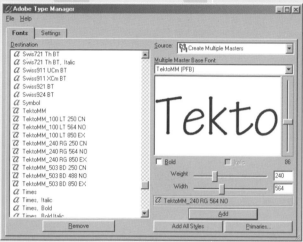

It was not long before designers were unhappy about seeing the still-jagged and rather imperfect bitmap type on their screens, even though they knew that the printed output would look good. Adobe then introduced Adobe Type Manager (ATM), which does away with the need for fixed-size bitmaps for screen use. Instead, ATM scales information from the PostScript outline font to create appropriately sized bitmap letterforms for the screen, for any size of type. Even so, on the Macintosh, ATM still needs at least one size of the bitmap fonts to be installed in order to display the font name in the menu. Type 1 fonts have now become a standard for digital software (ISO9541) and, worldwide, some 30,000 Type 1 fonts have been digitized for typesetting.

TRUETYPE

Despite the apparent dominance of Type 1 fonts, TrueType fonts have become very popular. They, too, are outline fonts, and their vector format allows them to be scaled infinitely with accurate results. Although the TrueType software standard was originally designed by Apple, TrueType fonts can now be used by both the Macintosh and Windows operating systems. Both systems have a TrueType rasterizer that provides information both for screen drawing and for print output. TrueType fonts are held in a single file (suitcase), which often contains sufficient information to construct the plain, plain italic, bold, and bold italic versions of the type family, making TrueType much tidier than the many files of Type 1 fonts.

MULTIPLE MASTER FONTS

Multiple Masters are Type 1 fonts originally developed by Adobe. They differ from other fonts in that they contain more than one digital outline for each character. For each character, a pair of outlines represents each end of a "design axis," and a Multiple Master font may contain axes for weight, width, style, or optical appearance—or all four together. With the Font Creator utility, the user can create custom fonts by making adjustments on a sliding scale between the extremes of each axis. Font Creator is built into ATM for Windows and is a stand-alone application on the Macintosh.

Font Creator draws information from the digital outline of each axis, resulting in a smooth morphing of shape and form, without distortion. Different widths and weights of type may be made by working between the width and weight axes. The style axis allows, for example, customized blending between, say, sans serif or serif versions or between slab serifs and bracketed serifs. The optical axis allows for change to

Tekto

Tekto

100 LT 250 CN

Tekto

503 BD 250 CN

Tekto

100 LT 564 NO

Tekto

503 BD 488 NO

Tekto

100 LT 850 EX

503 BD 850 EX

proportion, stroke weight, and spacing, which may all be adjusted to suit type size. This is similar to how the cutting of metal typefaces was varied among different sizes. In fact, a major criticism of much digital type has been the visual anomalies caused by proportional scaling from a single master. One set of proportions for stroke weight, width, fine detail, and spacing does not necessarily work well for all sizes.

Customized Multiple Master fonts thus allow for great flexibility of character shape without the kind of distortions associated with modifying horizontal scaling. Font files are bigger, though, and font names are complicated and difficult to keep track of. Multiple Master fonts are identified by MM plus a shortened version of the font name, e.g., ITCAvaGarMM. This is followed by two letters representing axes—BD for bold, LT for light, NO for normal, CN for condensed, EX for expanded, and OP for optical. A number indicates the axis value. Fonts you generated yourself will show these axes in lowercase.

Many type designers must feel that putting so much power into the hands of the end user defeats much of the effort put into carefully crafting a set of letterforms. Certainly, one of the strongest methods of emphasis available to designers is that of type contrast, and type designers have provided type styles and weights to support this. I believe that too much alteration and customization could well dilute typographic drama.

DETERMINING WHICH FONT FORMAT TO USE

There is no reason why one should not own both TrueType and Type 1 fonts. There is no problem in mixing them in a document, although it would be wise to make sure that you don't have fonts of the same name in the different formats, since that could lead to a lot of confusion. You do need to be sure that there is an outline font available to download to the printer, and it is worth checking the font folder to see that no fonts are missing. The example (bottom right) shows what you might expect to see in a typical font folder.

All TrueType fonts are kept in files called "suitcases," as are the bitmap fonts for Macintosh Type 1 fonts. PostScript fonts are two or more files kept loose in the Macintosh system Font folder or in the Windows PSFonts folder. An additional document, the font metrics file, is also kept there, for use by some software. This file has a set of directions called font metrics, which govern spacing, hinting, kerning, and kerning tables. Font vendors use differently designed icons to display their PostScript files, which can initially seem confusing.

Screen rendering of faux (false) bolds and faux italics are not identical. Where a family of font styles exist, the use of style menus to instruct for bold, italic, or bold italic is not recommended. The screen rendering may be close, but the printed outcome may not take up these attributes at all. Make a habit of selecting the TrueType style required through your font menu.

FAUX BOLD   BOLD

FAUX ITALIC

MAC TYPE 1   PC TYPE 1

Macintosh and PC Type 1 fonts. Macintosh users usually discard the AFM file, as it is only required by a few software packages.

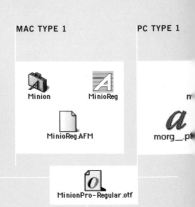

Minion   MinioReg

MinioReg.AFM   morg__p

MinionPro-Regular.otf

UNIVERSAL OPENTYPE

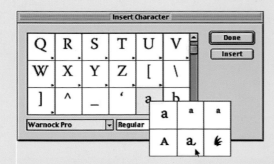

QuarkXPress's "font usage" dialog box, which lets the user keep track of fonts in the document. Invisible characters must be in one font or another, so fonts you don't remember using may occasionally show up in this window. It is good practice to find and change them to avoid needless queries from your print shop or service bureau.

ITALIC

Adobe InDesign's Insert Character dialog box is particularly useful for accessing unusual characters.

UNICODE FONTS

Most people regard the "name" and the "appearance" of a character as one and the same. In fact, the appearance or actual design of a letterform is called a "glyph." In terms of digital transmission, a character name is just a signal that says that an "a" falls here, without saying how it is drawn. Names become important in describing the set of characters in a font. Unicode is a relatively new standard for describing a "character set," working rather like the ASCII set. Most computers in the Western world use a character set of 200 characters. Unicode, however, will allow for some 65,000 separate characters. It has been devised for many reasons, the main one being for the better interchange of multilingual digital information. Characters may be rendered by mapping Unicode characters to glyphs. Unicode-compliant fonts may be designed to include glyphs accommodating a wide range of scripts that, on the one hand, share common letterforms and, on the other, have uniquely different ones. Common and composite glyphs may be made available on a single font. For example, the letter "a" is common in many Latin-alphabet scripts, so there is no need to provide more than one glyph design for "a" in a font style. An accented letter may be produced from two glyphs. Different glyph designs may be included within a font (e.g., swash, small cap, or lowercase "k") but they need only correspond to one character in the character set. Unicode is not in itself a font encoding device—fonts will be developed using the Unicode standard. How glyphs are displayed will depend on the rendering process.

OPENTYPE AND TRUETYPE OPEN

An exiting new development is the OpenType or TrueType Open font format, developed jointly by Microsoft and Adobe. The most obvious features are that there is no outward distinction between the two technologies and that they have true cross-platform capability. OpenType and TrueType Open fonts will consist of a single file, making installation and management easy.

абвгдеж
абвгдеж
αβγδεζη
αβγδεζη

An example of OpenType extra characters.

The stimulus for this development has been Unicode. The Unicode standard has been embraced by the major font developers, and this has allowed them to include characters in their fonts that had hitherto been contained in separate Expert Set fonts. These include ligatures(fi, fl, ffi, ffl, etc.), small capitals, extra accents and fractions, and other special characters.

Adobe InDesign already takes some advantage of OpenType, and other software will soon follow suit. In InDesign, special glyphs may be replaced by using the Insert Character dialog box.

WORLDTYPE

WorldType is Agfa Monotype's approach to providing extended character sets, and is also designed to conform to the Unicode standard. Whether the new font technology springing from the adoption of Unicode will be better coming from one developer or another is not clear. It is likely that any differences may be more significant to font designers and software developers than to the end user.

FONT PROBLEMS

Many font problems are self-inflicted, and a little organization and care should keep most people out of trouble. Probably one of the most common problems occurs at the print-out stage, e.g., a font is missing or seems not to be printing properly. A printing device must be able to access the outline font in order to print a typeface well. If you are printing directly from your own workstation using TrueType, the font outlines will be in a suitcase within the Fonts folder, and they should be found easily.

If you have used a Type 1 font, however, there is a possibility that the separate PostScript file for that font, or style of font, is missing or has been put into the wrong folder. The printer will then not find the correct outline data. In QuarkXPress and PageMaker it is possible to select a font style that has its own outline—Garamond Bold, for instance—and apply a bold style to it using the application's floating palette or keyboard. On screen, the type will look bolder than bold, but the printer will simply print a normal bold, using the font outline for Garamond Bold.

TrueType font suitcases usually contain information to create plain, bold, plain italic, and bold italic, and the style may be selected either through the drop-down font menus or by clicking the plain, bold, plain italic, and bold italic in a floating palette (see diagram). If a font has only one style (i.e., plain) in the suitcase, the system software will be able to display the four basic styles of

Background image is an example of a WorldType extended character set for Gujerati.

Shown below is TrueDoc Oriental.

plain, bold, plain italic, and bold italic on screen, so that the problem may not be apparent. When you print, no outline information will be found and the text will appear in the plain form but with incorrect metrics (spacing information).

It is wise to keep only one version of fonts that are identically or similarly named (e.g., TrueType AvantGarde and PostScript Avant Garde). It is very easy to mix them up, and they may not share the same metrics, which might easily disturb line endings and text flow. Be sure to send a copy of your font to the output bureau or print shop so that the type is rendered with the same outlines and metrics you used. You may do this legally only if the bureau has also licensed the font (*see* Font Copyright, below.)

Do not forget that blank lines and invisible characters such as spaces and returns also have font and style names attributed to them. It is advisable to make a final check of all blank lines and invisibles rather than have the printer or bureau calling for fonts that are "missing."

Character sets may differ between Macintosh and Windows (e.g., ligatures) and font metrics may not be identical. If your file originated on another platform, get further proofs. It is helpful to be able to recognize font icons, shown on page 32.

FONT COPYRIGHT

Font piracy is commonplace, but fonts are protected by copyright laws. When you "buy a font," you acquire in fact only a license to use it. Single-user licenses now allow only one copy of a font to be installed on one computer, rather than (as was once the case) a printer. Because of variations in font metrics, you need to have work output in exactly the font you used, so you may send a copy to a print bureau only if they hold a license for the same font. You may, of course, use the font to print from, as fonts are only temporarily downloaded in printing.

The TrueType font below contains plain (roman), italic, bold, and bold italic weights.

Font fine-tuning

Font fine-tuning

TRACK 1

TRACK 1

We have to start by assuming that we have chosen a type because, in its unaltered form, it meets our criteria for suitability of use in terms of mood and aesthetics, as well as all other criteria associated with the work in hand. Before we consider modifying, to any degree, the character spacing for any font, we need to be aware of its standard or default characteristics. We base our assessment on these standard values and, only then, decide on any modifications needed and how we may go about achieving the desired effects.

Most high-end page layout (DTP) and typesetting applications provide many built-in facilities for the control of character assembly. Because many aspects of spacing are interrelated, an alteration to one set of controls may well have a knock-on effect with others, which can initially be bewildering.

The safest way to understand what is going on is to examine and fully understand each of these controls in isolation. How each may interact with the others can then be better assessed.

HORIZONTAL SPACING CONTROL

Horizontal spacing is achieved through character and word spacing (*see* Part 2 for example). Typographers, typesetters, graphic designers, and publishers have always wanted some control over how character spacing behaves. Every type

SIDE BEARINGS

As the spacing between the two letters is produced purely by their side bearing, the letterspacing is considered to have a value of 0.

practitioner will have an individual view about how much is desirable. Despite the significant time, energy, attention to detail, and love that go into designing a type, it is virtually impossible for the typeface designer to cater for the countless variations of use, and the situations in which the type may be employed.

Tracking

Tracking is the term for the spacing between characters in a range or run of text. It should not be confused with kerning, which relates only to the adjustment of space between individual pairs of characters. We deal with kerning later but mention it here since some software uses the term "range kerning" for what most typographers understand as tracking.

Tracking may be applied in several ways, and how it is applied varies from one software package to another. Each character has a given width, determined by the type designer, which includes very small amounts of space on each side of the glyph, called the "side bearings." The side bearings ensure that, when text is set to the default, ideal standard envisaged by the type designer, the characters do not touch. When a range of characters is assembled normally, so that only the side bearings provide the white space between them, the tracking value is described as "0" (zero). We can depart from this norm by taking away a fractional space between each character or by adding it. This produces the effect of closer- or looser-spaced characters. It is because these adjustments involve tiny amounts of space that the old system of points and ems is still used by today's type designers, typographers, and computer systems. At approximately 0.35 mm (.01 inch), even one point is too large to be used

Laughter is the shortest distance between two people

TRACK –5

Laughter is the shortest distance between two people

TRACK 0 (NORMAL)

Laughter is the shortest distance between two people

TRACK +5

Laughter is the shortest distance between two people

TRACK +10

Laughter is the shortest distance between two people

TRACK +15

Laughter is the shortest distance between two people

TRACK +20

Victor Borge

as an increment of spacing for small sizes of type, so a much finer unit is derived from it. A little thought will also quickly show that for very large sizes of type one point might be too small an increment. What we need is a very small unit that varies relative to the type size. This unit is one-thousandth (0.001) of an em, where the em (unlike the pica em, which is always 12 points) is the same as the type size. In 10-pt type the em is 10 points, and all character spacing is based on thousandths of 10 points. Likewise, 7-pt type will use units of thousandths of 7 points. This is a microscopic unit, but as type size changes it remains optically constant. When tracking has been decreased, it is shown as a negative number (e.g., –7); when it is increased, it is shown as a positive number (e.g., 7). Some software uses 0.005 em as the standard increment instead, so you need to check what this number means in your software. For example, –3 in QuarkXPress is equivalent to Adobe InDesign's –15.

Tracking is global and (unless modified) is automatically set to a value of 0, i.e., the value designed into the font. Tracking can be changed automatically and globally, throughout a document, if desired. In this case, tracking values may be altered with a tracking editor (*see* example above). The new values will affect the font in question throughout the whole document.

Alternatively, or additionally, we may select sections of text as short as a word or as long as several pages, and manually apply localized tracking using menu commands. Remember that standard tracking evenly

Word spacing is unique to each typeface design and may vary from font to font.

I love deadlines. I especially like the whooshing
I love deadlines. I especially like the whooshing
sound they make as they go flying by.
sound they make as they go flying by.

Douglas Adams

ORIGINAL GARAMOND 16 POINT

GOUDY OLD STYLE 16 POINT

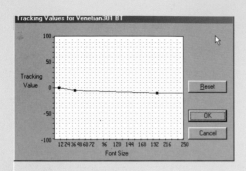

Example of one font, Goudy Old Style, with tracking values changing across a range of type sizes. As type becomes larger, the need to minus-track becomes progressively greater.

spaces all text of a given font according to basic font information. If we highlight sections of text and make localized changes to tracking, we must remember where we have done it; otherwise later alterations or amendments to the text may not carry our revised values.

The standard word space is a width determined by the font designer and is individual to each typeface, although in many cases it approximates to half or one-third of the em. In addition to varying in width from font to font, word spaces are also affected by changes in tracking values across a range of text.

Why would one want to alter tracking from the norm? When considering making changes to tracking, we should be clear that changes will be made for the right reasons. Regrettably, there is a bad, sloppy habit, adopted by many, of changing tracking for economic rather than esthetic reasons—an attempt to solve copy-fitting problems rather than a concern for the visual quality of the end result. To over-tighten tracking in order to squeeze large amounts of type into an area or to spread it out to fill a space is very tempting. This practice is used widely and indiscriminately, and is seen regularly in newspaper and magazine work. But it results in obviously contrived and unsightly text matter, and is especially noticeable where large areas of text have had tracking changed patchily, so producing ugly areas of unwarranted tonal change from paragraph to paragraph, column to column, or from one side of a spread to the other.

More desirably, tracking should be changed for esthetic reasons. Small type, for instance, while meeting the needs of the designer by other criteria, may begin to look too tight, resulting in diminished legibility. A small global increase to the tracking values here may well improve the overall look and ease of reading.

The discreet use of local tracking changes may be extremely useful for entirely practical purposes, e.g., where difficult pieces of copy may be gently tweaked in order to pull back a small, yet awkward, amount of "overmatter" or to make an

There is nothing permanent except ch

There is nothing permanent except change.
Heraclitus

VENETIAN 301 40 POINT TRACK –5

VENETIAN 301 10 POINT TRACK 0

There

VENETIAN 301 200 POINT TRACK –10

extra line to fill the column depth. Widow and orphan problems can often be solved by judicious changes to tracking over selected text. Adjustment to tracking can thus give valuable assistance, but only if used with care and sensitivity. It must be emphasized that great care needs to be exercised so that the tonal value of a range of type is not altered, or perceived to have been altered. The finesse with which type may be controlled allows for a great degree of subtle "cheating." So long as the cheating is not detectable, then I believe it is perfectly permissible.

The drawback to setting a single global tracking value for a font is that the revised tracking for one size of the type may be unsuitable for other sizes. For instance, where looser tracking may be appropriate for small text matter, a large headline in the same font may need to be tightened up. Fortunately, both QuarkXPress and Adobe PageMaker have tracking editors that allow tracking values to be adjusted automatically as sizes change. With the help of such editors, we may graphically plot a range of values that change appropriately with our choices of type size.

Tracking editors should be used when we wish to modify a font's standard tracking values throughout a document. Only then can we be confident that if new text is introduced by way of amendment or correction, our desired values will be maintained. If, on the other hand, we wish to vary tracking from one range of text to another to achieve a particular effect, we should manually apply different tracking to each selected area.

Kerning

The spacing between individual pairs of letters is called kerning. Whereas tracking values control a range of characters, kerning explicitly refers to the extra space placed uniquely between two specified characters.

Anyone who has had to draw letters by hand will be only too aware of the fact that some pairs of characters, when spaced out uniformly, do not always sit comfortably together, e.g., Wa, WA, Tr, and so on.

Soldiers of my Old Guard: I bid you farewell. For twenty years I have constantly accompanied you on the road to honor and glory. In these latter times, as in the days of our prosperity, you have invariably been models of courage and fidelity. With men such as you our cause could not be lost; but the war would have been interminable; it would have been civil war, and that would have entailed deeper misfortunes on France.

I have sacrificed all of my interests to those of the country.

I go, but you, my friends, will continue to serve France. Her happiness was my only thought. It will still be the object of my wishes. Do not regret my fate; if I have consented to survive, it is to serve your glory. I intend to write the history of the great achievements we have performed together. Adieu, my friends. Would I could press you all to my heart.

Soldiers of my Old Guard: I bid you farewell. For twenty years I have constantly accompanied you on the road to honor and glory. In these latter times, as in the days of our prosperity, you have invariably been models of courage and fidelity. With men such as you our cause could not be lost; but the war would have been interminable; it would have been civil war, and that would have entailed deeper misfortunes on France.

I have sacrificed all of my interests to those of the country.

I go, but you, my friends, will continue to serve France. Her happiness was my only thought. It will still be the object of my wishes. Do not regret my fate; if I have consented to survive, it is to serve your glory. I intend to write the history of the great achievements we have performed together. Adieu, my friends. Would I could press you all to my heart.

Napoleon Bonaparte

NEWS GOTHIC LIGHT 7.5/10 POINT TRACK –2

NEWS GOTHIC LIGHT 7.5/10 POINT TRACK 0

By minus-tracking a few lines or words, the odd word at the end of a paragraph can be brought back to make a more satisfactory ending. This very small amount of tracking will not affect the overall typographic texture of the page and is barely detectable.

s nothing

Font designers take these anomalies into account when designing type-faces by building kerning tables that work in the background, taking fractions of space away or adding it appropriately as letters are combined, resulting in more pleasing character relationships, e.g., Wa, WA, Tr, and so on (*see* the table opposite). It is important to note that, since kerning tables are built into the font metrics, any further manually applied kerning will be added or subtracted. For instance, when the cursor is placed between "W" and "A," a kerning value of 0 (zero) will be shown in the relevant control palette. This really means that there is neither more nor less space than that already provided for by the kerning table in the font metrics, e.g., –7. If you then subtract 10 units of kerning, it will be in addition to this existing value, i.e., –17.

In the setting of large or small amounts of copy, why and when should we start to intervene and set our own values? It is not advisable or practical to carry out any manual kerning to a body of text. The risk of missing similar kerned pairs is too great. In general, manual kerning is best applied only to display sizes.

As to altering kerning tables, I believe it takes a brave type user even to consider altering set values. Most type designers would be horrified at the thought of anyone tampering with elements of fine-tuning that had taken them months of hard work to perfect. There are, however, a few occasions when this might be forgiven. If the text contains a frequent occurrence of a particular pair, so that they appear more often than normal, modifying the values in the kerning table would allow a consistent solution if there was a problem in how the letters fit. For example, a frequently occurring trade name might contain a pair of characters that needs manual attention every time it is used. Fixing that pair by editing the kerning table will, however, alter the same pair in other words. It might also be that some plus or minus kerning in the kerning table does not match a very specific requirement, e.g., all numerals commonly have the same width (usually an en, or half an em) so that they will line up under each other in columns of figures. But if the number 11 or 111 appears in a flow of text, the numerals will seem too spaced out. If your document does not deal with numbers in columns, altering the kerning table of the 11 pair to a minus value will give a more satisfactory appearance. However, once this has been done in the kerning table, you will produce some strangely misaligned columns of figures.

H&Js (HYPHENATION AND JUSTIFICATION)

All the main typesetting and page-layout software packages offer methods by which the user may specify how and under what conditions automatic hyphenation will occur. They all, furthermore, offer the option of having no hyphenation at all, unless it is keyed in manually. Justification means that text is set so that both left and right line-endings range up in a given paragraph or run of paragraphs.

Different faces need different kerning treatment. Bracketed serifs may preclude some letters from being closed up too tightly, but sans serif letterforms will withstand considerable tightening up. Below certain user-definable text sizes, pair kerning can be so subtle that applications offer the option of turning auto kerning off.

-T	−7	F:	−7	OA	−6	T:	−19	WÅ	−17	y'	7
-V	−11	F;	−7	OV	−4	T;	−19	Wæ	−24	y"	7
-W	−11	FA	−13	OX	−5	TA	−11	Wø	−23	'A	−26
-X	−7	FS	4	OY	−4	Ta	−23	X-	−15	'J	−7
-Y	−15	FT	4	OÅ	−6	Tc	−25	XA	−13	'Å	−26
AC	−8	Fa	−5	P,	−41	Te	−24	XC	−6	Æ	−37
AG	−4	Fe	−5	P-	−15	Ti	−10	XO	−5	"A	−26
AO	−4	Fi	−7	P.	−41	To	−23	XT	−6	"J	−7
AQ	−7	Fo	−11	P:	−7	Tr	−24	Xe	−6	"Å	−26
AT	−15	Fr	−7	P;	−7	Ts	−25	XŒ	−5	"Æ	−37
AU	−6	Fu	−6	PA	−16	Tu	−26	XÅ	−13	ÅC	−8
AV	−17	Fy	−6	PU	−4	Tw	−31	XØ	−5	ÅG	−4
AW	−17	Fœ	−11	PY	−4	Ty	−31	Y,	−38	ÅO	−4
AX	−7	FÅ	−13	Pa	−7	Tœ	−23	Y-	−33	ÅQ	−7
AY	−18	Fæ	−5	Pe	−11	TÅ	−11	Y.	−38	ÅT	−15
Ac	−4	Fø	−11	Pn	−6	Tæ	−23	Y:	−32	ÅU	−6
Ad	−4	G-	4	Po	−10	Tø	−23	Y;	−32	ÅV	−17
Ae	−4	GA	−5	Pr	−4	UA	−11	YA	−21	ÅW	−17
Af	−4	GT	−7	Ps	−6	UÅ	−11	YC	−7	ÅX	−7
Ao	−8	GW	−6	Pœ	−10	V,	−41	YO	−7	ÅY	−18
Aq	−8	GY	−6	PÅ	−16	V-	−26	Ya	−28	Åc	−4
At	−4	GÅ	−5	Pæ	−7	V.	−41	Ye	−29	Åd	−4
Au	−4	JA	−5	Pø	−10	V:	−19	Yi	−11	Åe	−4
Av	−14	JÅ	−5	Q-	4	V;	−19	Yo	−31	Åf	−4
Aw	−9	K-	−7	Q'	4	VA	−17	Yu	−27	Åo	−8
Ay	−11	KA	−18	Q"	4	VO	−6	YŒ	−7	Åq	−8
AŒ	−4	KC	−10	R,	4	Va	−24	Yœ	−31	Åt	−4
A'	−26	KO	−12	R-	−4	Ve	−23	YÅ	−21	Åu	−4
A"	−26	KT	−6	R.	4	Vi	−13	YØ	−7	Åv	−14
Aœ	−8	KU	−8	RA	−11	Vo	−24	Yæ	−28	Åw	−9
AØ	−4	KW	−14	RT	−7	Vu	−19	Yø	−31	Åy	−11
Aø	−8	KY	−10	RV	−13	Vy	−15	f,	−7	ÅŒ	−4
B-	4	Ke	−4	RW	−17	VŒ	−6	f-	−4	Å'	−26
BS	5	Ko	−4	RY	−14	Vœ	−24	f.	−7	Å"	−26
BV	−6	Ku	−7	Ra	4	VÅ	−17	f'	26	Åœ	−8
BW	−9	Ky	−11	Ry	−6	VØ	−6	f"	26	ÅØ	−4
BY	−6	KŒ	−12	R'	−7	Væ	−24	r,	−11	Åø	−8
C-	5	Kœ	−4	R"	−7	Vø	−24	r-	−7	Đ-	7
CA	−4	KÅ	−18	RÅ	−11	W,	−42	r.	−11	ĐA	−7
CS	6	KØ	−12	Ræ	4	W-	−26	r'	7	ĐA	−7
C'	7	Kø	−4	SA	−5	W.	−42	r"	7	ĐW	−4
C"	7	LT	−15	SC	4	W:	−24	v,	−22	ĐY	−8
CÅ	−4	LU	−9	SG	4	W;	−24	v.	−22	ĐÅ	−7
D-	7	LV	−18	SO	4	WA	−17	v'	7	Ø,	−11
DA	−7	LW	−21	SQ	4	Wa	−24	v"	7	Ø-	4
DV	−7	LY	−17	SS	6	We	−23	w,	−22	Ø.	−11
DW	−4	Ly	−4	SŒ	4	Wi	−11	w.	−22	ØA	−6
DY	−8	L'	−30	SÅ	−5	Wo	−23	w'	4	ØV	−4
DÅ	−7	L"	−30	SØ	4	Wr	−20	w"	4	ØX	−5
F,	−20	O,	−11	T,	−22	Wu	−20	y,	−23	ØY	−4
F-	−4	O-	4	T-	−30	Wy	−18	y-	−4	ØÅ	−6
F.	−20	O.	−11	T.	−22	Wœ	−23	y.	−23		

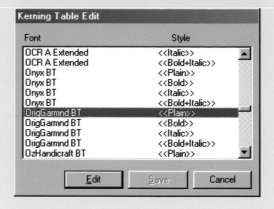

Original Garamond plain is selected to view a complete table of pair kerning values. Provided that auto kern is not disabled, the kerning table will automatically impose these values. It is interesting to note how much some pairs need to be drawn together, as shown by their large minus values. Note that tracking values are imposed over and above these pair kerning values.

Kerning is not necessary with numerals intended for tables or accountancy work, since numbers need to align vertically. However, kerning numerals for display or running text can be helpful.

111234

111234

The calculations used to assess break-points for hyphenation depend on the software knowing precise line lengths. Justification decisions are calculated on the basis of whether hyphenation is desired or not, and to what extent the spaces within a line may be stretched or compressed, so these two controls are very much interrelated. For this reason, typographers must consider hyphenation and justification as one single issue; this is why the term "H&J" is widely used to embrace these key aspects of typesetting. The designers of QuarkXPress have recognized this by displaying controls for both in a single dialog box called "H&Js." To see how these controls work, we will look at them in sequence, first separately and then together.

Justification

To understand justification, it is best to look first at what happens when type is set unjustified. As each character is keyed, it is placed after the previous one according to the standard spacing values contained in the font file. Each word is followed by the standard word space designed for that font. When a word is added to a line that is almost full, the software assesses whether there is room for it. If there is not, it will be forced onto the next line, leaving the line before slightly short. All this will take into account any kerning and tracking values that have been altered. The characters and their accompanying word spaces will look uniform and generally pleasing, but the right edge of the block of text will be ragged because the line lengths are uneven.

Justification aims to make all lines in a block of text end in the same place. As before, words and spaces are assembled on a line from left to right and when a word is added to a line that is almost full, the software assesses whether there is room for it. If there is not, the software quickly assesses two choices. Every word space on the line could be reduced to accommodate the word, or the word could be pushed onto the following line and the word spaces that are already on the line could be increased in order to fill the entire line length.

So that justification may be carried out successfully, the software uses certain routines in sequence to calculate the best choice. The best choice is that which provides the most uniform and satisfactory visual effect. Tightening up some word spaces may allow a word to be fitted on

---

"A free America, democratic in the sense that our forefathers intended it to be, means just this: individual freedom for all, rich or poor, or else this system of government we call democracy is only an expedient to enslave man to the machine and make him like it."
Frank Lloyd Wright

UNJUSTIFIED

"A free America, democratic in the sense that our forefathers intended it to be, means just this: individual freedom for all, rich or poor, or else this system of government we call democracy is only an expedient to enslave man to the machine and make him like it."
Frank Lloyd Wright

JUSTIFIED

Justification creates neat blocks of text but brings its own problems—particularly to narrow measures. Over-spacing and over-tightening inevitably occur in achieving evenly flush line ends. These can create "rivers" of white space running through the text, which affect readability. Hyphenation (word breaking) goes some way to alleviating this problem but is not always completely successful.

"A free America, demo-cratic in the sense that our forefathers intended it to be, means just this: individual freedom for all, rich or poor, or else this system of govern-ment we call democracy is only an expedient to enslave man to the machine and make him like it."
Frank Lloyd Wright

JUSTIFIED WITH HYPHENATION

Edit Hyphenation & Justification

Name:
Standard

☒ Auto Hyphenation
Smallest Word: 6
Minimum Before: 3
Minimum After: 2
☐ Break Capitalised Words

Hyphens in a Row: unlimited ▼
Hyphenation Zone: 0 mm

Justification Method
Min. Opt.
Space: 85% 110%
Char: 0% 0%

Flush Zone: 0 mm

☒ Single Word Justify

Cancel

"We hold these truths to be self-evident: that all men are created equal"

WORD SPACE 85% 110% 250% CHARACTER SPACE 0% 0% 4%

"We hold these truths to be self-evident: that all men are created equal"

WORD SPACE 85% 110% 250% CHARACTER SPACE –10% –5% 0%

"We hold these truths to be self-evident: that all men are created equal"

WORD SPACE 85% 110% 250% CHARACTER SPACE 5% 10% 15%

"We hold these truths to be self-evident: that all men are created equal"

WORD SPACE 80% 90% 100% CHARACTER SPACE 0% 0% 4%

"We hold these truths to be self-evident: that all men are created equal"

WORD SPACE 110% 130% 250% CHARACTER SPACE 0% 0% 4%

a line, but too much tightening will result in the words appearing to run together. So the software refers to a numerical limit below which word spacing may not be decreased; this limit is the *minimum word space*. Once this limit is reached, the software can only fit the word on the line if it reduces the space between characters. This is also subject to a limit: the *minimum character spacing*. If the software fails to fit it on the line, the word is pushed at last onto the next line. Reverting to the line it was working on, the software has now to increase the word spacing to spread the words evenly across the column width. Again, for the sake of esthetics, this increase is subject to a limit beyond which no more space may be added; this limit is the *maximum word space*. Once this limit is reached, the software can only fill out the line if it increases the space between characters. This is also subject to a limit: the *maximum character spacing*.

The whole process of justification depends on the software using minimum, "normal," and maximum values for spacing of both words and characters. (QuarkXPress calls these values "minimum," "optimal," and "maximum." I think "optimal" is a confusing term, because it in fact refers to the normal, unaltered, word space.) It must be remembered that, in determining any need for adjustment, the typesetting software addresses these values in sequence—first word spacing, followed by character spacing. Remember, however, that all calculations will take into account any change to kerning and tracking values you have made.

The whole objective of this process is to achieve lines of type that are justified (i.e., flush on both left and right) but do not appear to be unduly squashed or spread out, thereby producing an even and constant textural color. This is not always possible, and consequently there are situations where rows of unsatisfactory and ugly gaps, called "rivers," appear in a body of text, spoiling its overall appearance. This occurs most often when long words fall at the end of a line, and particularly when columns are narrow.

Hyphenation automatically follows the software's built-in recommended break point. These can be fine-tuned or overridden by setting values for the smallest breakable word and setting a minimum number of characters to fall before and after a break. The number of hyphens in a row may also be set. The hyphenation zone, used for unjustified text only, is measured from the right edge of the column. This feature can help control the raggedness of the line if required.

"If liberty and equal-
"If liberty and equality, as
"If liberty and equality, as is
ity, as is thought by
is thought by some, are
thought by some, are chiefly to
some, are chiefly to
chiefly to be found in
be found in democracy, they
be found in democra-
democracy, they will be
will be best attained when all
cy, they will be best
best attained when all per-
persons alike share in govern-
attained when all per-
sons alike share in government to the utmost."
sons alike share in
ment to the utmost."
government to the
utmost."
Aristotle

The same H&J settings used over three column widths. The widest measure has the most even texture and less hyphenation.

The settings here are:
word spacing:
98%, 100%, 110%
character spacing:
–1%, 2%, 4%

Hyphenation

All software provides facilities for hyphenation to break end-of-line words. Hyphenation may be used for justified or unjustified text, and the hyphens may be inserted manually or automatically.

UNJUSTIFIED TEXT

Hyphenation is normally applied to unjustified text only when it is ranged left. In visual terms, ranged-right and centered text do not lend themselves to hyphenation, although the rules of hyphenation will apply to those alignments, too.

Manual Hyphenation

When a long word is forced onto a new line leaving an unduly large space at the end of the line, manually typing a hyphen in the word will break it and allow the part before the hyphen to return and lengthen the preceding line. A danger of typing such a hyphen is that, if the text reflows owing to editorial or layout changes, the hyphen will travel with the word and may appear in the middle of a line. To avoid this problem, it is always advisable to type a "discretionary hyphen" (key command+hyphen in QuarkXPress, or control-click and choose Insert Special Character >Nonbreaking Hyphen in Adobe InDesign). A discretionary hyphen is an invisible command that instructs the software to break the word and insert a hyphen if the word happens to fall at the end of the line.

Note that the same keyboard commands, when entered with the cursor positioned at the start of a word, will prevent it from being hyphenated when auto hyphenation is switched on. This is useful if breaking a word (e.g., a name) is undesirable. Adobe InDesign calls this a nonbreaking command.

A hotel is a place that keeps the makers of 25-watt bulbs in business.

Shelly Berman

A hotel is a place that keeps the makers of 25-watt bulbs in business.

Shelly Berman

Some composite words are already hyphe... To ensure that they do not break and a... extra hyphen, it is best to key in a nonb... hyphen command *with the cursor place...* *start of a word*–key command+ - in Qua... control-click and choose Insert Special ... >Nonbreaking Hyphen in Adobe InDesi...

If a politician found he had cannibals among his many con-stituents, he would promise them missionaries for dinner.

H. L. Mencken

If a politician found he had cannibals among his many constituents, he would promise them missionaries for dinner.

H. L. Mencken

In the very short line shown above it is ... useful to insert a discretionary hyphen. This will break the word but should it appear elsewhere due to later editorial changes, the hyphen will disappear.

The English country gentleman galloping after a fox; the unspeakable in full pursuit of the uneatable.

Oscar Wilde

The English country gen-tleman gallop-ing after a fox; the unspeak-able in full pursuit of the uneatable.

Oscar Wilde

The English country gen-tleman galloping after a fox; the unspeakable in full pursuit of the uneatable.

Oscar Wilde

From left to right: no hyphenation, hyphenation on, hyphenation limited to one. Again a discretionary hyphen wou... work best here.

The charm of history and its enigmatic les-son consist in the fact that, from age to age, nothing changes and yet everything is completely dif-ferent.

Aldous Huxley

When auto hyphenation is on (*see above*), a Hyphenation Zone may be set (*see above right*). This has the effect of reducing the incidence of hyphenation but increasing the raggedness of the right-hand column edge.

The charm of history and its enigmatic lesson consist in the fact that, from age to age, nothing changes and yet everything is completely different.

10-MM HYPHENATION ZONE

Hyphenation looks more awkward in ranged-right and centered alignments.

Treat people as if they were what they ought to be, and you help them to become what they are capa-ble of being.

Johann Wolfgang von Goethe

Treat people as if they were what they ought to be, and you help them to become what they are capa-ble of being.

Johann Wolfgang von Goethe

Auto Hyphenation

When setting unjustified text with auto hyphenation turned on, a break will automatically be made in a long word that overflows the end of a line if that break will enable part of the word to fit on the line. The hyphen will be placed in accordance with either standard rules or modifications made to them. No hyphen will be added if you have inserted the nonbreaking command to prevent hyphen-ation. In unjustified text, the word and character spacing is not modified in any way from line to line.

Hyphenation Zone

The Hyphenation Zone permits further refinement in the process of auto hyphenation of unjustified text, by exerting some control over the raggedness of the right margin. When auto hyphenation is switched on but no Hyphenation Zone is specified, the last word in an unjustified line will be hyphenated according to the normal rules that determine word breaks if it is too long to fit. The result is often that the lines are of more even length, but also that they display an abundance of hyphens. If we were prepared to have less evenness in the line endings, we would need fewer hyphens. The Hyphenation Zone allows us to specify this choice. The zone is measured inward from the right side of the column, and its width in effect tells the software that no hyphenation is required as long as the line ends within the zone. If an unhyphen-ated word ends within the zone, the next word will not be broken, even if it could be hyphenated before the end of the line. If it is too long, it will simply be forced onto the next line. If a potentially hyphenated word coincides with the start of the Hyphenation Zone, or begins to its left, it will be broken. The use of Hyphenation Zones thus makes line lengths vary more (look more ragged) while reducing the incidence of hyphens. In unjustified typesetting, this feature provides an extremely useful method of controlling the shape of column edges.

JUSTIFIED TEXT

When type is justified, especially across a wide column, the general appearance of the type matter may look reasonably even, but there will always be occasions when the minimum and maximum values used by the justification process cannot deal adequately with long words that fall at the end of a line. Breaking the difficult word with a hyphen solves this problem.

Manual Hyphenation

The technique of manually inserting hyphens is the same as for unjustified text, and it is again advisable to use discretionary hyphens. Nonbreaking hyphens may be inserted in the same way.

Wide columns of type contain more word spaces and characters, and thus offer greater scope for the justification routines. In many cases, awkward words may be absorbed more readily without hyphenation. In other cases, however, at least some words will need to be broken. In narrow measures the need to hyphenate words becomes all the more necessary as opportunities for spreading or tightening are diminished. Although the hyphenation of a word does not unduly upset our ability to read, too many hyphenations mar the visual feel of a piece of text, so it is always the aim of a good designer to minimize the amount of hyphenation that occurs.

Auto Hyphenation
Automatic hyphenation is designed to produce the most satisfying results for most situations. As in the justification routines, hyphenation initially works according to the set of values that come as a default with the typesetting or page-layout software.

The software lets you set values for the minimum number of characters a word must contain for it to be auto hyphenated and for the number of characters allowed before and after the hyphen. You can also specify whether words starting with capital letters (e.g., names) may be auto hyphenated, and specify how many lines in sequence may end with a hyphen.

Hyphenation and Justification Working Together
Hyphenation and Justification processes interact. Hyphenation decisions, automatic or manual, are made only after the software has referred to the optimal, maximum, or minimum values, to check whether difficult words may be accommodated on a line to achieve a reasonable result. If this fails, the hyphenation process takes over—according to either manual input or auto hyphenation in accordance with the hyphenation values.

It is best to approach detailed typographic modification in a planned, controlled way. It is not, for example, useful to set up tracking until you know which typeface or column widths you will use.

Text set over a wide measure has more word spaces (and character spaces) that can be compressed or expanded to make a good line fit. Hyphenation will always help reduce the amount of exaggerated spacing, large or small.

HYPHENATION OFF

It is because modern education is so seldom inspired by a great hope that it so seldom achieves great results. The wish to preserve the past rather than the hope of creating the future dominates the minds of those who control the teaching of the young.

STANDARD H&J

It is because modern education is so seldom inspired by a great hope that it so seldom achieves great results. The wish to preserve the past rather than the hope of creating the future dominates the minds of those who control the teaching of the young.

Bertrand Russell

Elegance of language may not be in the power of all of us; but simplicity and straightforwardness are. Write much as you would speak; speak as you think.

If with your inferior, speak no coarser than usual; if with your superiors, no finer. Be what you say; and, within the rules of prudence, say what you are.

Alford

Elegance of language may not be in the power of all of us; but simplicity and straightforwardness are. Write much as you would speak; speak as you think.

If with your inferior, speak no coarser than usual; if with your superiors, no finer. Be what you say; and, within the rules of prudence, say what you are.

Alford

Assumed Basics
Typeface
Size
Weight
Column width
Alignment (ranged left or justified)

1 / First choice esthetic considerations
Auto kerning on / off
Tracking plus or minus
Horizontal scaling
(expanding or compressing letterforms)
Hyphenation on / off
Hanging punctuation yes / no

2 / Justified
Word spacing
 – minimum
 – optimal (preferred standard)
 – maximum
Character spacing
(Optimal prevails unless word spacing values fail)
 – minimum
 – optimal (preferred standard)
 – maximum

3 / Unjustified
Word spacing
 – only optimal needs to be set
Character spacing
 – only optimal needs to be set

4 / Hyphenation on
Controls
Minimum characters before hyphen
Minimum characters after hyphen
Minimum word size
Break capitalized words on /off
Number of hyphens in a row
Hyphenation Zone
(for unjustified text only)

It is tempting to think that this automated decision-making means that we don't need to worry how hyphenation or justification works. This is not so. There are too many variables in typographic design—too many idiosyncratic demands arising from text assembly, and too many conditions under which finished work may be viewed—for any standard values to work consistently well. Poor hyphenation inevitably gives text an unsatisfactory appearance. The more control we have over word-breaking, the better our chance of producing even text matter.

WHEN AND WHERE TO TAKE CONTROL

In considering adjustments to hyphenation and justification values, the designer must have in mind the sort of result that is being sought. "Suitability for the job" states the requirement neatly, but interpreting "suitability" can be a hard task. Before hyphenation and justification are fine-tuned, more fundamental issues should have been resolved—choice of typefaces, appropriate type sizes and styles, together with desired column widths. It is important to understand that default or standard typographic values for all typesetting software have been set to suit average setting situations, i.e., using average-sized type to an average column width, using an average-looking typeface. But when was anything average? And when did you ever want to create an average piece of text or display setting? The truth of the matter is that layout, projected message, and audience, among many other factors, are rarely average. What we need is to take hold of our typesetting software and make it work brilliantly for us. Jonathan Hoefler, of

Hoefler Type Foundry, is quoted as having said, "When you're designing a typeface, you are really making a product. You're building a machine that will go to make other products." I believe that we can, with care, take this "machine" and make it work for us while maintaining its intrinsic integrity. If this means getting our hands dirty by modifying H&Js, so be it.

In this discussion of typographic fine-tuning, I am putting aside for now those aspects of handling type that may be intended to serve decorative or emotive purposes or, maybe, to shock. In these cases, total manual control over every word and character may be required, and relevant kerning, tracking, and hyphenation choices will probably depend on a wholly unique set of esthetic values. And we talk about display setting later.

What we are trying to do here is to control our textsetting so that we may achieve a body of text, whether it be a paragraph or a dozen pages, that meets page-layout requirements in terms of texture, color, shape, cohesion, harmony, readability, and, in the end, has the desired impact.

The overall color and texture of a body of type and its general appearance are governed by horizontal spacing, i.e., word and character spacing, and by vertical spacing, i.e., line spacing (leading) and paragraph spacing. Vertical and horizontal spacing are interrelated. For instance, characters spaced densely along the line will look black on the page. Open out the lines by increasing leading, and the overall tonal value lightens. Which comes first? The chicken or the egg?

In order to address so many variables—tracking, automatic kerning, manual kerning, justification, hyphenation, and line and paragraph spacing, not to mention the choice of type, size, and weight—one needs to adopt a systematic approach to these issues. It is useful to have in the back of one's mind some of the fundamental and general effects of the decisions we make about our typography; we need to be at least roughly aware of the results likely to come from our changes in specifications.

Deciding on typeface and size has to be the first consideration, as all aspects of setting values relate directly to them. Choice of typeface is determined by the mood and tone that you wish to create. I have referred to the printed word as "visible speech." There are innumerable kinds of "voice"—soft, loud, firm, commanding, persuasive, kindly, seductive—and regional accents add extra characteristics. All these, provided they are given the opportunity and space to make a good delivery, are capable of imparting a clear, strong message. As with speech, so also with type: the range of types for a printed message is equally wide. Weight of type determines the stridency of the voice, and size determines its volume. Understanding the project and the message to be conveyed is key to being able to select a suitable typeface.

The diagram below is a general guide to the decision-making process H&J software. Established first are basic metrics such as type size, style, user-defined kerning, auto-kerning, and tracking information.

When the last word of a line is typed, is the line longer than the specified column width? If yes, Route A is followed.

Making Changes

Listed below are some rule-of-thumb results or effects that can be expected from text set to particular specifications.

- In fully justified setting, wide measures throw up fewer problems than narrow measures. The more word spaces there are in a line, the less they will each have to be reduced or enlarged to allow the line of type to be justified.
- Whereas wide measures reduce the need to hyphenate, narrow measures not only increase the number of words needing to be hyphenated but also increase the risk of sequential lines having hyphens at the end.
- Reduced tracking darkens the perceived tonal value of text on a page. Characters are closer to each other, producing a denser texture.
- Unhyphenated justified text will produce a greater number of inconsistencies in tonal color from line to line. Without the facility to hyphenate, the software must solve short and over-long line endings with word- and character-spacing alone, which usually results in the occurrence of occasional pale, over-stretched lines, and dark, compressed lines through the body of text. The narrower the column, the more often this happens.
- Unhyphenated, ranged-left text will produce a more exaggerated raggedness on the right margin.
- Narrow measures of justified text are more likely to produce white rivers.
- Excessive latitude in minimum and maximum word space values tends to exaggerate the amount of word space relative to the words themselves. A form of clumping tends to occur.
- Little latitude in minimum and maximum word space values combined with too much latitude in character spacing tends to produce words that appear to run into one another.

Tracking

These are the sorts of circumstance in which to consider changing tracking values.

- A slight lightening of the textural color is required: increasing overall tracking will produce more white space between characters.
- A slight darkening of the textural color is wanted: decreasing overall tracking will reduce the amount of white space between characters.
- It is discovered that an unusual range of problem words cannot adequately be accommodated by the use of hyphenation alone: a very small change in tracking over several lines may bring back or push forward offending matter—but this technique should be used only if it is not detectable to the eye.
- More characters need to be fitted into a given column width, but changes to type or type size are not deemed a better alternative.

Justification

Controlling how justification works is probably the most important tool that the typographer has to ensure that bodies of text maintain some reasonable consistency of color and texture. It is also an important tool for minimizing end-of-line

H&J OFF

Those parts of the system that you can hit with a hammer (not advised) are called hardware; those program instructions that you can only curse at are called software.
Author unknown

H&J STANDARD

Those parts of the system that you can hit with a hammer (not advised) are called hardware; those program instructions that you can only curse at are called software.
Author unknown

H&J OFF WITH MANUAL LINE-BY-LINE TRACKING

+4 Those parts of the sys-
+10 tem that you can hit
+11 with a hammer (not
+11 advised) are called
+8 hardware; those pro-
+5 gram instructions that
+9 you can only curse at
+6 are called software.
Author unknown

Plus- or minus-tracking values, line-by-line, can greatly enhance narrow columns. This is practical only in small amounts of text, such as are used in advertising and promotional work, as it is labor-intensive.

problems. To understand what happens when justification values are modified, it is useful to see what happens to text when the default values are in place. Remember that "standard" or average values suit an average typeface set in a column of average width.

Minimum, optimal, and maximum values determine the limits governing the spaces between words and between characters. The software refers to these limits in order to achieve justification.

- The minimum word space value refers to the minimum amount to which space between words may be decreased in order to achieve justification.
- The optimal word space value refers to the space between words that will be used as the actual fixed word space in unjustified text. It is also the initial value against which the software makes its calculations to achieve justified lines. Percentage relates to a percentage of the word-space size set in the font file, which differs from font to font.
- The maximum word space value refers to the maximum amount to which space between words may be increased in order to achieve justification.

Minimum, optimal, and maximum values also apply to character spacing. The percentage still means a percentage of the character spacings supplied in the font file, but some software (e.g., QuarkXPress) treats the percentage as space added or taken away, so that the supplied value is 0 (zero), while other software treats the supplied value as 100%. Minimum and optimal character-space values are by default usually set to the supplied value, i.e., 0 or 100%, with only the maximum value different, e.g., 4 or 104%. These default settings may seem conservative, but they are set like this because character spacing does not usually need either to be increased much, or decreased at all, if justification is achieved by adjusting word spacing first. Only when justification fails on all attempts to vary word spacing will the software try to vary character space.

The default values work best in average conditions to suit average needs. Innovative graphic designers and typographers find that few of their setting needs are average and off-the-shelf setting values aren't always the best.

Consider changing the default value if:
- The default settings produce unsatisfactory hyphenation.
- You want to permit more character spacing rather than word spacing.
- You want to permit more word spacing rather than character spacing.
- You need to resolve problems caused by a high incidence of long or short words.
- You want to minimize the amount of hyphenation.
- You prefer to increase the amount of hyphenation to maintain better text color.

WHAT IS THE EFFECT OF MAKING H&J CHANGES?

The following is a guide to the effect of H&J changes:

- As the difference between minimum and maximum word-space values increases, so does the risk of holes and rivers occurring in the text.
- When great decreases and increases are applied to minimum and maximum character-space values, variations in textural color from line to line will be more marked.

H&J 50/70/100
"They think they can make fuel from horse manure. Now, I don't know if your car will be able to get 30 miles to the gallon, but it's sure gonna put a stop to siphoning."
Billie Holliday

STANDARD H&J 85/110/250
"They think they can make fuel from horse manure. Now, I don't know if your car will be able to get 30 miles to the gallon, but it's sure gonna put a stop to siphoning."
Billie Holliday

H&J 30/100/500
"They think they can make fuel from horse manure. Now, I don't know if your car will be able to get 30 miles to the gallon, but it's sure gonna put a stop to siphoning."
Billie Holliday

The examples above show the effects of different minimum, optimal, and maximum word spacing. Despite a difference of 400 in the maximum word spacing between the top and bottom examples, the resulting text setting looks almost the same.

- Decreases to optimal word or character spacing (or both) will make the overall tonal color of the text darker. Increases will make it lighter.
- Changes to minimum and maximum values will have greater impact in narrow columns than in wide ones—remembering, of course, that the perceived width of a column is relative to type size. It is because of this relativity that values relating to changes to word and character spacing are always expressed in percentage (i.e., relative) terms.

Multi-line Composition

A general drawback to justification is that—owing to the unpredictability of how words fall and the way in which software treats justification line by line—there is a tendency for some lines of type to be too tight (squashed) and some to be too loose (opened up). Careful attention to word and character spacing values will help to reduce these difficulties. However, the multi-line composer—an Adobe innovation, in its InDesign software—helps to solve this problem very successfully. The software considers several lines of type together, assigns penalty points to each line, and then calculates the best way to handle spacing and hyphenation to smooth out irregularities from line to line.

A tip

Is your column width what you think it is? A QuarkXPress text box has a default text inset value of 1 point. You can change it to 0 point.

VERTICAL SPACING

The color and texture of your text, modified by use of horizontal word and character spacing, will also be affected by your choice of vertical line spacing. As lines of type are moved apart, more white space on the page is revealed, thus creating an overall lightening of the body of text. By the same token as lines get closer, so the text becomes denser and darker. Clearly some good control over how this is achieved would be useful.

The examples below show how a change in tracking values alone, using a standard set of H&J values, will provide different line ending results.

- STANDARD H&J TRACK +5
- STANDARD H&J TRACK 0
- STANDARD H&J TRACK −5

Hardly a year passes that fails to find a new, oft-times exotic, research method or technique added to the armamentarium of political inquiry. Anyone who cannot negotiate Chi squares, assess randomization, statistical significance, and standard deviations is less than illiterate; he is preconscious.

A. James Gregor

Leading

Line space is controlled by leading values, usually expressed in points. In modern usage, the term "leading" means the distance from the baseline of one line of type to the baseline of the next, which includes both the type and any extra space added (the old usage meant only the extra space). Some software allows you to specify a mode of leading, either "word-processing" or "typographic"; always choose "typographic." In most software, the default value for leading is "auto," which usually has a value of 120% of the type size. For 10-pt type, "auto" leading will thus produce baselines that are 12-pt apart (written 10/12, and pronounced "ten on twelve"). It is often said that most text faces need a few points of leading (note the old usage, meaning extra space between lines), and in average text settings "auto" leading will always produce readable text. Again, few jobs are average, and "auto" leading will often introduce too much space between lines, making the block of text look pale and uninteresting. It is thus best to take control over line space and alter the value in your software's preferences to 100%; you will then know its absolute value. Alternatively, you could make it a habit to replace "auto" leading immediately with a definite value. You might start with leading equal to the type size ("set solid"), and then assess the effect of increasing the value in, say, 0.25 point increments.

Leading may be applied to text locally as required, but that risks irregularities. Consistency is as vital to typographic spacing as to well-modulated speech. Text that hiccoughs along in unregulated bursts is bound to distract the reader, diminishing the effectiveness of the message, so controlling vertical spacing in an automated way is a useful tool for the designer.

Be sure you know what auto leading means. It often has a default value of 120%, i.e., 20% extra leading over and above the type size. It is best to go to your preferences and set auto to 100%. The word "auto" will still appear but you will know it equals your type size and you can modify it quickly.

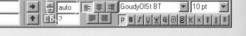

Leading has a great impact on the tonal color of text on a page. If, for instance, letter and word spacing appear too open, an increase in leading will help to even out the texture.

A human being is a part of the whole, called by us Universe, a part limited in time and space. He experiences himself, his thoughts and feelings as something separated from the rest, a kind of optical delusion of his consciousness. This delusion is a kind of prison, restricting us to our personal desires and to affection for a few persons nearest to us. Our task must be to free ourselves from this prison by widening our circle of compassion to embrace all living creatures and the whole of nature in its beauty.
Albert Einstein

10/10 POINT (SET SOLID)

A human being is a part of the whole, called by us Universe, a part limited in time and space. He experiences himself, his thoughts and feelings as something separated from the rest, a kind of optical delusion of his consciousness. This delusion is a kind of prison, restricting us to our personal desires and to affection for a few persons nearest to us. Our task must be to free ourselves from this prison by widening our circle of compassion to embrace all living creatures and the whole of nature in its beauty.
Albert Einstein

10/12 POINT (AUTO)

A human being is a part of the whole, called by us Universe, a part limited in time and space. He experiences himself, his thoughts and feelings as something separated from the rest, a kind of optical delusion of his consciousness. This delusion is a kind of prison, restricting us to our personal desires and to affection for a few persons nearest to us. Our task must be to free ourselves from this prison by widening our circle of compassion to embrace all living creatures and the whole of nature in its beauty.
Albert Einstein

10/14 POINT

Humanity needs practical men, who get the most out of their work, and, without forgetting the general good, safeguard their own interests. But humanity also needs dreamers, for whom the disinterested development of an enterprise is so captivating that it becomes impossible for them to devote their care to their own material profit.

Without doubt, these dreamers do not deserve wealth, because they do not desire it. Even so, a well-organized society should assure to such workers the efficient means of accomplishing their task, in a life freed from material care and freely consecrated to research.

Marie Curie

Paragraph Attributes

Formats | Tabs | Rules

Left Indent:	3.5 mm
First Line:	-3.5 mm
Right Indent:	0 mm
Leading:	9.5 pt
Space Before:	0 mm
Space After:	0 mm
Alignment:	Left
H&J:	Standard

☐ Drop Caps
Character Count: 1
Line Count: 3

☐ Keep Lines Together
◉ All Lines in ¶
◉ Start: 2 End: 2

☐ Keep with Next ¶
☒ Lock to Baseline Grid

Apply   Cancel   OK

"Space After" is a useful way of controlling and editing paragraph spacing globally without having to key in extra returns to provide white space. If a "Space Before" value is also given, it will be added to any "Space After." "Space Before" is usually best kept for headings, where additional space is required to set it off farther from the previous paragraph.

Paragraph Spacing

If you want paragraph spacing, you could just add a blank line (i.e., key two returns). But blank lines or multiple blank lines are not always appropriate. Using paragraph formatting lets you specify any value as extra space between paragraphs (*see* dialog box above).

Baseline Grids

Many purists like the baselines of type in one column of text to align accurately with the baselines in neighboring columns. This contributes greatly to the beauty of good text setting, but it is not always easy to accomplish. As headings, cross-headings, and paragraph spaces, all of different sizes, are introduced into a run of text, the baselines in one column can soon get out of step with those in the next. This is where the baseline grid and the "snap to" feature comes in. Most software lets you set a grid of equally spaced horizontal lines to which all your baselines can be snapped; you can specify how far down the page the grid starts, and how far apart the gridlines should be. If you apply "snap to guides" to your text, all the baselines will be forced to align with the grid, even if you have given different paragraphs different leading values. With grid increment and leading equal, you will see no response if you reduce the leading; the type is constrained to the grid. Equally, you will see too much response if you increase the leading; the type will snap to alternate gridlines. To escape this straitjacket, you could apply "snap to guides" to some paragraphs and not others—but then why bother with a baseline grid at all? Why introduce irregularity to a grid that was meant to give regularity? It might be better to design typographic detailing around the regularity of the incremental size of the baseline grid. If we decide that 10-pt text in our chosen type should have 11-pt leading, paragraphs could be 10/11, major headings could be 24/33, and subheads 14/22, with space after paragraphs and extra space before headings also specified in 11-pt increments. With picture boxes also dimensioned in multiples of 11 point, it will be seen that the whole page structure, in vertical terms, would create satisfying horizontal stresses and be devoid of unsightly stepping from column to column. Without compromising text line alignment or the grid, we can also use the "baseline shift"

This diagram shows how different leading values will cause baselines to differ across columns.

Thomas Jefferson

To Jean Nicholas Demeunier, 1786

"What a stupendous, what an incomprehensible machine is man! Who can endure toil, famine, stripes, imprisonment & death itself in vindication of his own liberty, and the next moment inflict on his fellow men a bondage, one hour of which is fraught with more misery than ages of that which he rose in rebellion to oppose."

To Congress, 1806

"I agree with you that it is the duty of every good citizen to use all the opportunities, which occur to him, for preserving documents relating to the history of our country."

To John Wyche, 1809

"I have often thought that nothing would do more extensive good at small expense than the establishment of a small circulating library in every county, to consist of a few well-chosen books, to be lent to the people of the country under regulations as would secure their safe return in due time."

To Miles King, 1814

"Our particular principles of religion are a subject of accountability to our god alone. I enquire after no man's and trouble none with mine; nor is it given to us in this life to know whether yours or mine, our friend's or our foe's, are exactly the right."

The major heading and the subheads, though different in size from the 11 pt text, have leading values that are multiples of 11 pt and therefore sit exactly on the underlying baseline grid.

The subhead below has a baseline shift of +3.5 point, which raises it without altering the integrity of the baseline grid.

"What a stupendous, what an incomprehensible machine is man! Who can endure toil, famine, stripes, imprisonment & death itself in vindication of his own liberty, and the next moment inflict on his fellow men a bondage, one hour of which is fraught with more misery than ages of that which he rose in rebellion to oppose."

To Congress, 1806

"I agree with you that it is the duty of every good citizen to use all the opportunities, which occur to him, for preserving documents relating to the history of our country."

command to float our headings flexibly where we wish to see them (*see* picture). Ultimately grids should serve our needs; we are not slaves to the grid.

Baseline Shift

Baseline shift is an extremely useful way of handling small vertical spacing problems. Despite its benefits, there are times when mathematical regularity is just not good enough, e.g., when a full line of capitals falls between two lines of upper- and lowercase letters, or when capitals are vertically centered in a text box.

Style Sheets

Leading and paragraph spacing may be incorporated into a style sheet, along with all other aspects of control, including tracking, baseline shift, and H&Js. By using styles, you can ensure that all aspects of your setting, including both horizontal and vertical spacing, remain constant and consistent. Style sheets are probably the most effective and productive way of controlling your work.

Putting It All Together

Controlling all these aspects to achieve your desired typography may seem daunting. So many aspects of control are interrelated, and changing one detail of any produces knock-on effects. Thus, your hyphenation method is influenced, not just by its own variables, but by tracking, kerning, word and character spacing, all of which in their turn are influenced by whether the text is justified or not. And so it goes on, each piece of the typographical jigsaw affecting all the others.

Abcdef	Abcdef
Abcdef	Abcdef
Abcdef	Abcdef
Abcdef	Abcdef
Abcdef	Abcdef

abc	abc
Abcdef	Abcdef
Abcdef	Abcdef
Abcdef	Abcdef
Abcdef	Abcdef
Abcdef	Abcdef

This shows schematically how, on the left, a 15/18 subhead pushes subsequent lines out of alignment whereas, on the right, "lock to baseline grid" maintains consistency of line feed.

WIDOW

mental.

You got to be careful if you don't know where you're going, because you might not get there.

No wonder nobody comes here – it's too crowded.

You can observe a lot by watching.

Ninety percent of the game is half mental.

Yogi Berra

ORPHAN

Widows and orphans—"widow" is the term for the last line of a paragraph that falls at the top of a column or page. An "orphan" is the first line of a paragraph that falls at the base of a column or page, while the rest of the paragraph is carried over. (One-word lines at the end of a paragraph are also sometimes called "widows.")

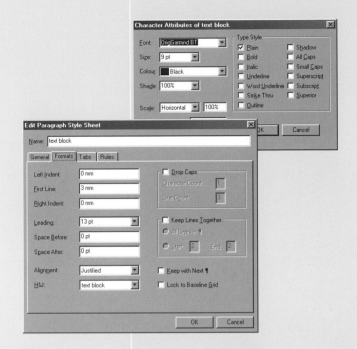

Take each area of possible control one step at a time. Everyone will have his or her own way of working, but set out below is an ordered approach to making decisions about type:

1 Choice of typeface and type sizes for text and display will be greatly influenced by client, readership, creative assessment, and the physical constraint of being able to put this much information into that much space. You are likely to get this information early on.

2 Column width will be dictated by layout considerations, which may include, among other issues, the relationship of type and image, the way in which the copy has been written and needs to be parceled up and interpreted, the degree of indenting that may be necessary, and the opportunities offered by the format.

3 Once choices of type and column width are made, the color of text on the page will be most influenced by choices of leading and any tracking. Wide leading accentuates the horizontal stress of each line of type; positive tracking tends to reduce horizontal stress.

4 Paragraph shape is influenced by the decision whether to justify or not, and also by the decision whether to hyphenate or not. If automatic hyphenation is wanted, Hyphenation Zones will help to control the raggedness of right margins and thus paragraph shapes.

5 Regularity and harmony are the final, yet very important, fine-tuning issues. These considerations will be essential to achieving cohesion and helping to delineate the underpinning structure. The smoothing out of textural irregularities, such as holes and rivers, and poor texture will require attention to word and character spacing.

6 Display work, more often than not, requires localized fine-tuning to kerning pairs. But these considerations, although essential to the overall quality of the typography, are less urgent to resolve at an early stage. There are many who choose and craft their display type with meticulous care only to accompany it with sloppy-looking text.

Lack of care over any element of typographic content can so easily upset the mood, balance, and elegance of the whole printed page.

It is highly recommended that style sheets are created for all aspects and values of typographic style. This ensures that complex instructions are applied consistently throughout a document.

HANGING PUNCTUATION

Punctuation characters normally fall inside the column width of set text. When they are at the start or end of a line, they can sometimes create the appearance of ugly indents in the text and disrupt an otherwise uniform margin.

In extensive amounts of relatively small running text, these indents will not be so apparent that they give cause for concern. But when small amounts of textural matter or semi-display work form a visual focal point, steps need to be taken to correct problems caused by end-of-line punctuation. Adobe InDesign provides an efficient, automatic hanging punctuation feature, which Adobe calls "optical margins." Most punctuation characters that appear at the start or end of a line, such as start quotes, end quotes, hyphens, periods, and commas, will be pushed away from the flush edge of type. Colons and semicolons, in view of their shape, are not hung.

Hanging punctuation can be achieved in both QuarkXPress and Adobe PageMaker, but it requires a workround and effort to get good results; the technique relies on the use of left and right indents in the paragraph format specifications. The neat aspect of InDesign's automatic hanging punctuation is that it allows text to reflow without further attention.

LEGIBILITY AND READABILITY

What is legible might not be easy to read. An obvious demonstration of this would be single characters that are arranged vertically to form a shop sign. The individual letters are perfectly legible but the complete word or name is difficult to read.

There are naturally occasions when graphic energy and excitement may be created with words and letterforms, when readability is less important overall than decoration, mood, or the dramatic effect that the assembled letterforms create. However, much of our communication is through printed words that need to be rapidly absorbed and understood, with little effort required from the reader. Once reading becomes hard work, readership is lost.

A little understanding of how text is read will help us to produce effective, readable communication. The eye scans information and passes it to the brain for interpretation. The speed and ease with which the brain is able to process the information will have a significant bearing not only on comprehension but also on the ability to maintain high levels of concentration. Nothing lessens the strain and helps the brain to work better than order and structure. The less we tire the reader, the more likely the

" Democracy means simply the bludgeoning of the people by the people for the people."

" Experience is the name that everyone gives to their mistakes."

" Fashion is a form of ugliness so intolerable that we have to alter it every six months."

" For those who like that sort of thing, it's the sort of thing that they like."

" I hope you have not been leading a double life, pretending to be wicked and being really good all the time. That would be hypocrisy."

" I sometimes think that God, in creating man, overestimated His ability."

Oscar Wilde

Hanging punctuation is created in QuarkXpress by manually inserting values to the Left Indent and First Line settings in the Format dialog box. Adobe InDesign has an optical margin feature that creates hanging punctuation automatically.

There are, of course, times when typographic intrigue and visual excitement will take precedence over legibility.

ONYX

FEAR

ENGLISCHE SCHREIBSCHRIFT

Relax

SCIENCE

LUCIAN

Journal

RUGGED

Each typeface, whether for display or text, has a unique feel that will influence the effect of word groupings and consequently readability.

VENETIAN 301

CALM

FRANKLIN GOTHIC HEAVY

WIN

reader is to understand the message and keep reading for longer—which is exactly our objective.

Written language is understood by eye and brain scanning swiftly across a page. Rather than consciously noting individual characters, the eye-brain mechanism processes groupings of characters (words) and the shape of those groupings (phrases). Anything we do to promote the quick recognition of such groupings will help the reader to absorb information quickly and comfortably.

Choice of Type

The design of a type will have an immediate influence on how well word groupings work. The ability and speed at which groupings can be assimilated will vary from one reader to another, but the following observations will always be true.

Line Length or Column Width

As the eye comes to the end of a line, it attempts, in an instant, to find the start of the next line. If the line is too long, the eye will have to travel too far, and one of two things will happen: by the time the eye finds the start of the line there has been a perceptible interruption to the flow of comprehension or, worse, the wrong line gets picked up. If the line is too short, it will contain so little information that phrases and comprehension become disjointed. The eye-brain mechanism will also tire from the combined effort of finding the next line while remembering the content of the last. Harmonious flow in word and phrase recognition depends heavily on the designer considering line length. The best length is widely accepted as around 60 characters or 10 to 12 words.

Kerning and Tracking

Clear, distinct word formation is paramount for ease of reading. Spacing characters too closely or openly marginally diminishes the reader's ability to recognize word shapes. Multiplied hundreds of times, this marginal effect can strain the eye and brain. No single distraction is likely to annoy or slow a reader, but an accumulation of hundreds of tiny faults inevitably makes for hard reading.

Leading

Poor line spacing produces problems similar to those caused by poor line length. If lines are spaced out excessively, the eye takes a fraction longer to move down the page and, if this is uncomfortable, tiredness again creeps in. Spacing lines too closely increases the possibility of their being skipped and tempts the eye to stray.

There is a
theory
which states
that if ever
anybody dis-
covers
exactly what
the Universe
is for and
why it is
here, it will
instantly dis-
appear and
be replaced
by some-
thing even
more
bizarre and
inexplicable.
There is
another the-
ory which
states that
this has
already hap-
pened.
Douglas
Adams

Dynamic Letter Groups

Since words are seen and understood as groups or phrases, it is best to avoid, wherever possible, weakening their integrity. When lines of type are spaced too tightly, it is easy for the smoothness of linear reading to be jostled up and down by the touching, or near touching, of ascenders and descenders. Furthermore, nothing more reliably destroys the elegance of a group of words than automatic underlining, which weakens the relationship of adjacent characters and also cuts right through their descenders. Ultimately, it is the care given to these considerations that will determine how long a reader's attention may be kept.

The foregoing comments are benchmarks that provide for the most sustainable and comfortable reading. They form a basis from which we can diverge to suit our own projects. What may be considered typographically appropriate for small bursts of text and display work might be wholly inappropriate for a 300-page novel. A treatment that might suit a broadsheet might not suit a magazine.

The point to bear in mind—always—is that the decisions we make about typeface and type spacing will have an impact not only on the legibility of text but also on its readability.

Layout and Real Color

The scope of this book does not allow for either an exhaustive investigation into layout techniques or the use of color (colored inks as distinct from the textural color of type on the page). Nevertheless, it is worth noting that all the detailing of type and spacing we have discussed will be greatly affected by layout and by colors of ink used. Light inks, for instance, will quieten down text that may, in black, look aggressive and unappealing. Small type can be given importance by dynamic disposition. Contrasting one textural element with another may produce subtle changes in each. Good typography depends on a host of interrelated visual elements, and none may be ignored.

In summary, it would be fair to say that it is the designer or typographer's job to successfully use a font to its fullest and best advantage by understanding, selecting, and using the fine tuning techniques available.

There is a theory which states that if ever anybody discovers exactly what the Universe is for and why it is here, it will instantly disappear and be replaced by something even more bizarre and inexplicable. There is another theory which states that this has already happened.
Douglas Adams

Eddington, SIR ARTHUR
1882–1944

Proof is the idol before whom the pure mathematician tortures himself.

In N. Rose, Mathematical Maxims and Minims, 1988.

We used to think that if we knew one, we knew two, because one and one are two. We are finding that we must learn a great deal more about 'and'.

In N. Rose, Mathematical Maxims and Minims, 1988.

We have found a strange footprint on the shores of the unknown. We have devised profound theories, one after another, to account for its origins. At last, we have succeeded in reconstructing the creature that made the footprint. And lo! It is our own.

Space, Time and Gravitation, 1920.

I believe there are 15,747,724,136,275,002, 577,605,653,961,181,555,468,044,717,914,527, 116,709,366,231,425,076,185,631,031,296 protons in the universe and the same number of electrons.

The Philosophy of Physical Science. Cambridge, 1939.

Eigen, MANFRED
1927–

A theory has only the alternative of being right or wrong. A model has a third possibility: it may be right, but irrelevant.

Jagdish Mehra (ed.), The Physicist's Conception of Nature, 1973.

Einstein, ALBERT
1879–1955

This has been done elegantly by Minkowski; but chalk is cheaper than grey matter, and we will do it as it comes.

J. E. Littlewood, A Mathematician's Miscellany, 1953.

Common sense is the collection of prejudices acquired by age eighteen.

In E. T. Bell, Mathematics, Queen and Servant of the Sciences. 1952.

Everything should be made as simple as possible, but not simpler.

Reader's Digest. Oct. 1977.

I don't believe in mathematics.

Albert Einstein, Quoted by Carl Seelig.

Imagination is more important than knowledge.

On Science.

The most beautiful thing we can experience is the mysterious. It is the source of all true art and science.

What I Believe.

God does not care about our mathematical difficulties. He integrates empirically.

L. Infeld, Quest, 1942.

(About Newton)
Nature to him was an open book, whose letters he could read without effort.

In G. Simmons, Calculus Gems, 1992.

As far as the laws of mathematics refer to reality, they are not certain; and as far as they are certain, they do not refer to reality.

In J. R. Newman (ed.), The World of Mathematics, 1956.

What is this frog and mouse battle among the mathematicians?

In H. Eves, Mathematical Circles Squared, 1972.

These thoughts did not come in any verbal formulation. I rarely think in words at all. A thought comes, and I may try to express it in words afterward.

In H. Eves, Mathematical Circles Adieu, 1977.

An example of fairly complex text, including headings, subheadings, and comments, given structure and order by careful leading, paragraph spacing, and baseline shift.

1 14-pt Garamond Bold
2 12-pt Garamond Regular Expert SC
3 12-pt Garamond Regular Expert SC, nonlining numerals, with a –2.5-pt baseline shift
4 Leading 13-pt throughout
5 Text 12-pt Garamond Regular
6 8-pt Garamond Italic
7 Typographer's punctuation
8 First line baselines may be set to ensure that first lines of type will align across columns, even if they are of different point sizes.
9 Equal column depth

abcdefghijklmnop

PART 2

SLIMMING

fine fonts

Individual letters make words

and words make up sentences

(or lines) and lines make up

paragraphs. Paragraphs make up

columns of text and columns of

text make up pages. When

handling type we are dealing

with designed letterforms.

letters

sentences

paragraphs

columns

pages

letterforms

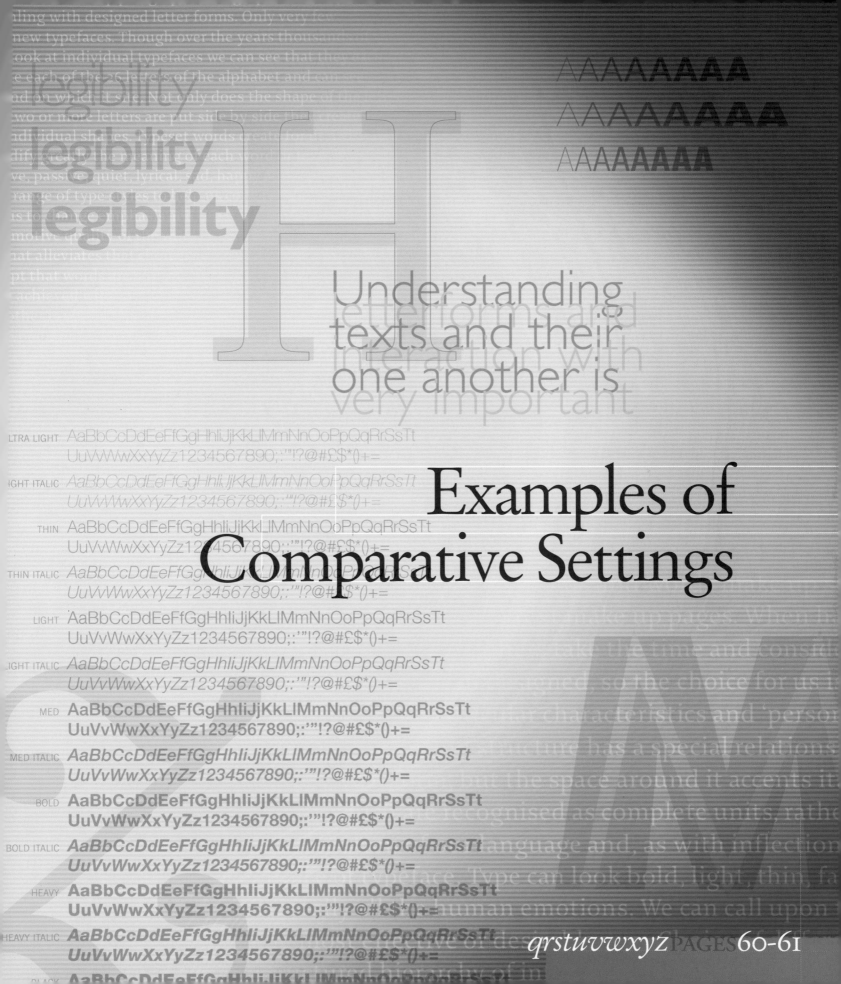

Understanding
texts and their
one another is
very important

Examples of
Comparative Settings

ULTRA LIGHT AaBbCcDdEeFfGgHhIiJjKkLlMmNnOoPpQqRrSsTt
UuVvWwXxYyZz1234567890;:'""!?@#£$*()+=

LIGHT ITALIC *AaBbCcDdEeFfGgHhIiJjKkLlMmNnOoPpQqRrSsTt*
UuVvWwXxYyZz1234567890;:'""!?@#£$()+=*

THIN AaBbCcDdEeFfGgHhIiJjKkLlMmNnOoPpQqRrSsTt
UuVvWwXxYyZz1234567890;:'""!?@#£$*()+=

THIN ITALIC *AaBbCcDdEeFfGgHhIiJjKkLlMmNnOoPpQqRrSsTt*
UuVvWwXxYyZz1234567890;:'""!?@#£$()+=*

LIGHT AaBbCcDdEeFfGgHhIiJjKkLlMmNnOoPpQqRrSsTt
UuVvWwXxYyZz1234567890;:'""!?@#£$*()+=

LIGHT ITALIC *AaBbCcDdEeFfGgHhIiJjKkLlMmNnOoPpQqRrSsTt*
UuVvWwXxYyZz1234567890;:'""!?@#£$()+=*

MED **AaBbCcDdEeFfGgHhIiJjKkLlMmNnOoPpQqRrSsTt**
UuVvWwXxYyZz1234567890;:'""!?@#£$*()+=

MED ITALIC ***AaBbCcDdEeFfGgHhIiJjKkLlMmNnOoPpQqRrSsTt***
UuVvWwXxYyZz1234567890;:'""!?@#£$*()+=

BOLD **AaBbCcDdEeFfGgHhIiJjKkLlMmNnOoPpQqRrSsTt**
UuVvWwXxYyZz1234567890;:'""!?@#£$*()+=

BOLD ITALIC ***AaBbCcDdEeFfGgHhIiJjKkLlMmNnOoPpQqRrSsTt***
UuVvWwXxYyZz1234567890;:'""!?@#£$*()+=

HEAVY **AaBbCcDdEeFfGgHhIiJjKkLlMmNnOoPpQqRrSsTt**
UuVvWwXxYyZz1234567890;:'""!?@#£$*()+=

HEAVY ITALIC ***AaBbCcDdEeFfGgHhIiJjKkLlMmNnOoPpQqRrSsTt***
UuVvWwXxYyZz1234567890;:'""!?@#£$*()+=

qrstuvwxyz

Controlling text

Controlling text

The following section shows comparative examples of text set in thirty popular classic and contemporary typefaces in a variety of weights.

All the examples are displayed in 9-pt type with additional leading of 20%, which is the default value or "factory setting" in most page layout applications. The longer examples are settings of average line length (between 10 and 12 words on a line), and the shorter examples show how working with narrower columns affects the appearance of the block of type. There are both fully justified settings, in which both left and right margins are made to align, and unjustified settings, in which the right margin is ragged. (Unjustified setting is also called "left-justified" or "ragged right.")

Settings in the top row use QuarkXPress default Hyphenation and Justification values for Space (word spacing) and Character

(character spacing). Auto hyphenation is switched on in some examples (the factory setting) and switched off in others in order to show the effect on the word spacing. With hyphenation turned on, the maximum number of hyphens in a row is two. Most applications can apply tracking in increments of 0.001 em, where the em is taken to be twice the width of the 0 (zero) in the size of the typeface being used, i.e., its value is relative. Although 0.001 em can be specified in QuarkXPress, the basic increment employed is $\frac{1}{200}$ em (0.005). Thus tracking or kerning of plus 4, is equal to 0.02 em. You should, therefore, check the increments and definition of the em for your own software.

The other examples vary from typeface to typeface and show how adjusting spacing through the Hyphenation and Justification facility of your software can improve the appearance of setting, especially fully justified setting. You can set three values each for word spacing and character spacing, and these values interact with each other when the software decides where to end a line (and thus how much space to add between words and characters in fully justified setting). The variations may therefore seem endless. However, there are some straightforward benchmarks you should bear in mind when adjusting the values.

First, in addition to the shape of the glyphs, fonts contain information about normal (i.e., 100%) spacing values for both characters and words, and may also contain information used for automatic (or background) kerning of character pairs. When you consider fine-tuning, bear in mind that the type designer will have lavished great effort on crafting these spatial relationships, which determine the overall look of a block of text. Second, the DTP software comes with factory-set default values for hyphenation and justification (H&Js), which are applied over and above what the type designer specified as normal. Factory-set defaults often produce reasonably even text setting for average contexts, so there may be no need to alter these values, although house style may determine different criteria for word breaks. So most of us will start from the benchmark of the software's default H&Js. It is when you depart from the average column width or tracking that adjustment to word and character spacing values may be necessary.

Decisions you make about horizontal spacing values are also influenced by vertical spacing, i.e., leading. Tighter leading will have the effect of exaggerating variations in word or character

1, 5, 6, 9, 10 These examples are all set with QuarkXPress default H&J values across a wide (**1**) and narrow (**5, 6, 9, 10**) measure.

5, 6 Show justified type set with hyphenation on (**5**) and off (**6**).

9, 10 Type set ragged right with hyphenation on, but with the Hyphenation Zone set to 0mm (**9**) and 14mm (**10**).

2, 3, 4 Shows text set to a wide measure with alternative H&J and

tracking values designed to improve the overall appearance of text setting in the featured font. Example **4** uses the values most suitable for setting with hyphenation turned off.

7, 8 Show text set to a narrow measure with alternative H&J values designed to improve the overall appearance of text setting.

11 Shows unjustified text set with optimal H&J values, with hyphenation turned off.

HYPHENATION AND JUSTIFICATION DEFAULTS					
WORD SPACING			CHARACTER SPACING		
MIN	OPT	MAX	MIN	OPT	MAX
QUARK	85%	110%	250%	0%	0%	4%
PAGEMAKER	80%	100%	110%	−1%	0%	1%
INDESIGN	80%	100%	133%	0%	0%	0%
FREEHAND	80%	100%	150%	−10%	0%	10%
ILLUSTRATOR	100%	100%	200%	0%	0%	5%

spacing. If large spaces are unavoidable, increasing the leading may help to hide the problem. Tighter leading may even suggest that you apply slightly tighter tracking to obtain evenness of typographic color.

With so many variables, different typefaces set to the same width will throw up different spacing problems and places at which words may or may not hyphenate. Though the following examples are comparative, they can in no way be definitive, so you would be well advised to carry out your own comparisons to ensure your complete satisfaction with the overall tone, texture, and readability of your text setting. Examples of how text faces work in display sizes are also shown on selected pages.

JUSTIFIED TEXT

In general, wide columns present fewer problems than narrow ones, since longer lines offer more opportunity for large end-of-line words to be absorbed without undue squeezing. With more words per line, the amount by which each word space needs to be increased can be fairly minimal when a big word is pushed over onto a new line. It follows that in a long line the difference between minimum and maximum values for word spaces need not be great. Too large a difference will allow some lines to be crowded or "tight," with narrow word spaces, while others will be "loose," with unduly large spaces between words.

Specifying a rather narrow range of values for word spacing will ensure greater evenness in texture. If hyphenation is also used in conjunction with this narrow range, it may be a good idea to set the maximum character space at a slightly higher value to keep an even texture.

More problems occur in fully justifying narrow columns, since shorter lines will contain fewer words and thus offer the software fewer opportunities to even out line lengths by adjusting word spacing. This can result in more lines being spaced very tight (with minimum word spacing) and others being spaced very loose (with maximum word spacing), which will disturb the rhythm of the text. To minimize this problem, it can help to reduce the word

spacing values and the difference between them. This may result in a few tight lines but will help to prevent unsightly gaps. Hyphenation can help to reduce these gaps further.

Given that in narrow columns there will inevitably be a little letterspace compression on many lines, it might well be worth considering applying a small amount of plus-tracking to the text overall to alleviate the effect, as well as using automatic hyphenation. When word spacing values are fairly tight, a small amount of minus-tracking can help to provide a more even texture.

Remember that column width is not just a measurement on paper but is also affected by the typeface. A generous column width for one typeface might be too narrow for another, owing to differing character widths.

Note that Flush Zone (not to be confused with Hyphenation Zone) forces the last line of a paragraph to fill the column width even if it falls a little short of that width. The zone is measured from the right edge of the column. If the final character of the paragraph falls into this area, the line will be justified. It is not wise to use a large Flush Zone, since ugly stretching could occur, but a small Flush Zone can prove useful for situations in which a squared-off paragraph shape is required.

UNJUSTIFIED SETTING

With unjustified type, minimum and maximum word spaces are irrelevant. It is the optimal value that is important, since this controls the constant width of all word spaces. The factory or default settings of your software may increase word spacing to a higher value than either the type designer originally intended (i.e., to more than 100%) or you expected. Optimal character spacing may also be set by your software's factory defaults to something other than zero. You may need to adjust these values for yourself.

The Hyphenation Zone is an often misunderstood facility. Many typographers will switch off hyphenation when setting ranged left text as hyphenation tends to erode the raggedness inherent in unjustified type. You are able to increase this raggedness, and at the same time decrease the incidence of word breaking, by increasing the Hyphenation Zone. The zone is measured from the right edge of the column. A word that has an acceptable hyphenation point will break (be hyphenated automatically) only if the preceding word falls short of the zone.

Key

W ⬚	= WORD SPACE
C ⬚	= CHARACTER SPACE
H ✓	= HYPHENATION ON
T ✓	= TRACKING ON
H	= HYPHENATION OFF
T	= TRACKING OFF
HZ	= HYPHENATION ZONE

Bembo

Bembo is a successful revival of Francesco Griffo's original directed by Stanley Morison, the type designer and printing historian. Bembo belongs to the group of typefaces known as Old Face that became the standard text type of the 16th and 17th centuries in Europe. The original cut was used by Aldus Manutius to print Cardinal Bembo's *De Aetna*, a treatise on his visit to Mount Etna, hence its name.

Issued by The Monotype Corporation in 1929, Bembo is a classical, evenly designed text face with well-proportioned characters that continues to survive the centuries.

TYPE CHARACTERISTICS

The Bembo family includes regular (roman), semibold, bold, and extra bold, each with italic versions, while Bembo Expert features special characters. It is an extremely good text face with a quietly elegant character and a traditional feel. Its relatively small x-height, but open form, makes for economy of space yet provides excellent legibility, suitable for a wide range of purposes and subject matter, particularly books. Curiously there are two versions of the italics: Narrow Bembo Italic, which was designed by Alfred Fairbanks, the calligrapher, and independently issued as Bembo Condensed Italic, and Bembo Italic, still with a condensed feel but more open and distinctively angular.

Bembo has an angled stress to the bowls of its characters, a straight bar to the lowercase "e" and oblique serifs to the lowercase letters. The capital letters are slightly shorter than the ascenders of the lowercase letters.

There are distinctly recognizable characteristics to the following capital letters: the "K" has bowed strokes, the "R" a wide apart junction and sweeping downstroke, the "T" divergent serifs to the crossbar, and the "W" has crossed center stokes with a flat base. In the lowercase letters, the "a" has a small bowl, the "j" has a long curled tail and the leftmost stroke of the "m" and "n" angles slightly inward giving a faintly bowed effect.

ABCDEFGHIJKLMNOPQRSTUVWXYZ
abcdefghijklmnopqrstuvwxyz1234567890;:'"!?@#£$&★()+=

Quark Default H&J

	Min	Opt	Max
W ✐	85	110	250
C ✐	0	0	4
H ✓	ON		
T	0		

Individual letters make words and words make up sentences (or lines) and lines make up paragraphs. Paragraphs make up columns of text and columns of text make up pages. When handling type we are dealing with designed letterforms. Only very few experienced typographers take the time and considerable effort to design new typefaces. Though over the years thousands of typefaces have been designed, so the choice for us is enormous. When we look at individual typefaces we can see that they each have their own particular characteristics and "personality." We may examine each of the 26 letters of the *alphabet and can see that their form and structure has a special relationship with the background on which it sits. Not only does the shape of the letter itself give it form, but the space around it accents its indi-*

User Defined H&J

	Min	Opt	Max
W ✐	90	100	105
C ✐	0	0	3
H ✓	ON		
T ✓	3		

Individual letters make words and words make up sentences (or lines) and lines make up paragraphs. Paragraphs make up columns of text and columns of text make up pages. When handling type we are dealing with designed letterforms. Only very few experienced typographers take the time and considerable effort to design new typefaces. Though over the years thousands of typefaces have been designed, so the choice for us is enormous. When we look at individual typefaces we can see that they each have their own particular characteristics and "personality." We may examine each of the *26 letters of the alphabet and can see that their form and structure has a special relationship with the background on which it sits. Not only does the shape of the letter itself give it form, but the space*

User Defined H&J

	Min	Opt	Max
W ✐	90	100	120
C ✐	0	3	10
H ✓	ON		
T	0		

Individual letters make words and words make up sentences (or lines) and lines make up paragraphs. Paragraphs make up columns of text and columns of text make up pages. When handling type we are dealing with designed letterforms. Only very few experienced typographers take the time and considerable effort to design new typefaces. Though over the years thousands of typefaces have been designed, so the choice for us is enormous. When we look at individual typefaces we can see that they each have their own particular characteristics and "personality." We may examine each of the 26 letters of *the alphabet and can see that their form and structure has a special relationship with the background on which it sits. Not only does the shape of the letter itself give it form, but the space around it accents its*

Quark Default H&J	Min	Opt	Max
W	85	110	250
C	0	0	4
H ✓	ON		
T	0		

Individual letters make words and words make up sentences (or lines) and lines make up paragraphs. Paragraphs make up columns of text and columns of text make up pages. When handling type we are dealing with designed letterforms. Only very few experienced typographers take the time and considerable *effort to design new typefaces. Though over the years thousands of typefaces have been designed, so the choice for*

Quark Default H&J	Min	Opt	Max
W	85	110	250
C	0	0	4
H	OFF		
T	0		

Individual letters make words and words make up sentences (or lines) and lines make up paragraphs. Paragraphs make up columns of text and columns of text make up pages. When handling type we are dealing with designed letterforms. Only very few experienced typographers take the time and *considerable effort to design new typefaces. Though over the years thousands of typefaces have been*

Quark Default H&J	Min	Opt	Max
W	85	110	250
C	0	0	4
H ✓	ON		
T	0		

Individual letters make words and words make up sentences (or lines) and lines make up paragraphs. Paragraphs make up columns of text and columns of text make up pages. When handling type we are dealing with designed letterforms. Only very few experienced typographers take the time and considerable *effort to design new typefaces. Though over the years thousands of typefaces have been designed, so the*

Quark Default H&J	Min	Opt	Max
W	85	110	250
C	0	0	4
H ✓	ON HZ 14mm		
T	0		

Individual letters make words and words make up sentences (or lines) and lines make up paragraphs. Paragraphs make up columns of text and columns of text make up pages. When handling type we are dealing with designed letterforms. Only very few experienced typographers take the time and *considerable effort to design new typefaces. Though over the years thousands of typefaces have been*

User Defined H&J	Min	Opt	Max
W	75	100	110
C	-1	-1	3
H ✓	ON		
T	0		

Individual letters make words and words make up sentences (or lines) and lines make up paragraphs. Paragraphs make up columns of text and columns of text make up pages. When handling type we are dealing with designed letterforms. Only very few experienced typographers take the time and considerable *effort to design new typefaces. Though over the years thousands of typefaces have been designed, so the choice for*

User Defined H&J	Min	Opt	Max
W	75	100	115
C	0	0	10
H ✓	ON		
T ✓	2		

Individual letters make words and words make up sentences (or lines) and lines make up paragraphs. Paragraphs make up columns of text and columns of text make up pages. When handling type we are dealing with designed letterforms. Only very few experienced typographers take the time and considerable *effort to design new typefaces. Though over the years thousands of typefaces have been designed, so the*

User Defined H&J	Min	Opt	Max
W	85	100	250
C	0	0	0
H	OFF		
T	0		

Individual letters make words and words make up sentences (or lines) and lines make up paragraphs. Paragraphs make up columns of text and columns of text make up pages. When handling type we are dealing with designed letterforms. Only very few experienced typographers take the time and considerable effort to design new typefaces. Though over the years thousands of typefaces have been designed, so the choice for us is enormous. When we look at individual typefaces we can see that they each have their own particular characteristics and "personality." We may examine each of the 26 letters of the alphabet and can see that their form and structure has a special relationship with *the background on which it sits. Not only does the shape of the letter itself give it form, but the space around it accents its individuality. When two or more letters are put side by side they create words which are recognised as complete*

User Defined H&J	Min	Opt	Max
W	90	100	105
C	0	0	3
H	OFF		
T	0		

Individual letters make words and words make up sentences (or lines) and lines make up paragraphs. Paragraphs make up columns of text and columns of text make up pages. When handling type we are dealing with designed letterforms. Only very few experienced typographers take the time and considerable effort to design new typefaces. Though over the years thousands of typefaces have been designed, so the choice for us is enormous. When we look at individual typefaces we can see that they each have their own particular characteristics and "personality." We may examine each of the 26 letters of the *alphabet and can see that their form and structure has a special relationship with the background on which it sits. Not only does the shape of the letter itself give it form, but the space around it accents its*

ITALIC
*AaBbCcDdEeFfGgHhIiJjKkLlMmNnOoPpQqRrSsTt
UuVvWwXxYyZz1234567890;:'"!?@#£$&★()+=*

SEMIBOLD
**AaBbCcDdEeFfGgHhIiJjKkLlMmNnOoPpQqRrSsTt
UuVvWwXxYyZz1234567890;:'"!?@#£$&★()+=**

SEMIBOLD ITALIC
***AaBbCcDdEeFfGgHhIiJjKkLlMmNnOoPpQqRrSsTt
UuVvWwXxYyZz1234567890;:'"!?@#£$&★()+=***

BOLD
**AaBbCcDdEeFfGgHhIiJjKkLlMmNnOoPpQqRrSsTt
UuVvWwXxYyZz1234567890;:'"!?@#£$&★()+=**

BOLD ITALIC
***AaBbCcDdEeFfGgHhIiJjKkLlMmNnOoPpQqRrSsTt
UuVvWwXxYyZz1234567890;:'"!?@#£$&★()+=***

EXTRA BOLD
**AaBbCcDdEeFfGgHhIiJjKkLlMmNnOoPpQqRrSsTt
UuVvWwXxYyZz1234567890;:'"!?@#£$&★()+=**

EXTRA BOLD ITALIC
***AaBbCcDdEeFfGgHhIiJjKkLlMmNnOoPpQqRrSsTt
UuVvWwXxYyZz1234567890;:'"!?@#£$&★()+=***

Bodoni (Bauer)

Bodoni, designed in the late 18th century by the celebrated Italian printer, Giambattista Bodoni, is a typeface of both fine and grand design. His typeface, with its precise, reasoned feel and dramatic contrast between thick and thin strokes, has a pronounced verticality that departs entirely from any calligraphic influence. The result is a beautifully measured, elegant, and crisp design that responds to generous space. Bodoni is one of the best-known examples of the so-called modern faces, although not an everyday design. There have been many revival versions of Bodoni that, although embracing the essence of the original design, have modified and reduced the extreme thick and thin contrast. The text settings on these spreads are set in Bauer Bodoni.

TYPE CHARACTERISTICS

Bodoni has typically extreme thick and thin strokes, minimally bracketed serifs, and the hairline horizontal foot serifs on the lowercase letters. The long ascenders and descenders give it a graceful quality.

Recognizable characteristics in the capital letters are defined top and bottom serifs on the "C," a generously seriffed low bar to the "G" (as on the original design), rather narrow "M" and "W," a "Q" with a low tail that has a vertical start to it, and crossed-stroke "W" with a hint of space between the first and center serifs.

In the Bauer lowercase letters, the "c," "f," and "r" have ball finishes. The "g" has a small upper bowl by comparison to the generous lower one. The top serifs of Bauer's "b," "h," and "l" are slightly concave. Also, unlike other Bodoni versions, Bauer's "y" has an almost unique bent stroke and subtle curl to the tail, and the ampersand has a rhythm of form that echoes the lowercase letterforms. The single downstroke of the dollar sign is clean and uncluttered.

The Bauer Bodoni family consists of regular and bold fonts, with italics; small caps and Old Style figures; a black weight with corresponding italics; and a condensed bold and black version.

ABCDEFGHIJKLMNOPQRSTUVWXYZ
abcdefghijklmnopqrstuvwxyz1234567890.,:;'"!?@#£$&*()+=
ABCDEFGHIJKLMNOPQRSTUVWXYZ1234567890

Quark Default H&J

		Min	Opt	Max
W	✎	85	110	250
C	✎	0	0	4
H	✓	ON		
T		0		

Individual letters make words and words make up sentences (or lines) and lines make up paragraphs. Paragraphs make up columns of text and columns of text make up pages. When handling type we are dealing with designed letterforms. Only very few experienced typographers take the time and considerable effort to design new typefaces. Though over the years thousands of typefaces have been designed, so the choice for us is enormous. When we look at individual typefaces we can see that they each have their own particular characteristics and *"personality." We may examine each of the 26 letters of the alphabet and can see that their form and structure has a special relationship with the background on which it sits.*

User Defined H&J

		Min	Opt	Max
W	✎	90	100	100
C	✎	0	0	3
H	✓	ON		
T	✓	3		

Individual letters make words and words make up sentences (or lines) and lines make up paragraphs. Paragraphs make up columns of text and columns of text make up pages. When handling type we are dealing with designed letterforms. Only very few experienced typographers take the time and considerable effort to design new typefaces. Though over the years thousands of typefaces have been designed, so the choice for us is enormous. When we look at individual typefaces we can see that they each have their own particular characteristics and *"personality." We may examine each of the 26 letters of the alphabet and can see that their form and structure has a special relationship with the back-*

User Defined H&J

		Min	Opt	Max
W	✎	90	100	120
C	✎	0	3	10
H	✓	ON		
T		0		

Individual letters make words and words make up sentences (or lines) and lines make up paragraphs. Paragraphs make up columns of text and columns of text make up pages. When handling type we are dealing with designed letterforms. Only very few experienced typographers take the time and considerable effort to design new typefaces. Though over the years thousands of typefaces have been designed, so the choice for us is enormous. When we look at individual typefaces we can see that they each have their own particular characteristics and *"personality." We may examine each of the 26 letters of the alphabet and can see that their form and structure has a special relationship with the background on*

Quark Default H&J

	Min	Opt	Max
W ✐	85	110	250
C ✐	0	0	4
H ✓	ON		
T	0		

Individual letters make words and words make up sentences (or lines) and lines make up paragraphs. Paragraphs make up columns of text and columns of text make up pages. When handling type we are dealing with designed letterforms. Only very few experienced typographers take *the time and considerable effort to design new typefaces. Though over the years thou-*

Quark Default H&J

	Min	Opt	Max
W ✐	85	110	250
C ✐	0	0	4
H	OFF		
T	0		

Individual letters make words and words make up sentences (or lines) and lines make up paragraphs. Paragraphs make up columns of text and columns of text make up pages. When handling type we are dealing with designed letterforms. Only very few experienced typographers take *the time and considerable effort to design new typefaces. Though over the years*

Quark Default H&J

	Min	Opt	Max
W ✐	85	110	250
C ✐	0	0	4
H ✓	ON		
T	0		

Individual letters make words and words make up sentences (or lines) and lines make up paragraphs. Paragraphs make up columns of text and columns of text make up pages. When handling type we are dealing with designed letterforms. Only very few experienced typographers *take the time and consider-able effort to design new typefaces. Though over the*

Quark Default H&J

	Min	Opt	Max
W ✐	85	110	250
C ✐	0	0	4
H ✓	ON	HZ 14mm	
T	0		

Individual letters make words and words make up sentences (or lines) and lines make up paragraphs. Paragraphs make up columns of text and columns of text make up pages. When handling type we are dealing with designed letterforms. Only very few experienced typographers *take the time and consider-able effort to design new typefaces. Though over the*

User Defined H&J

	Min	Opt	Max
W ✐	75	90	90
C ✐	-3	0	3
H ✓	ON		
T	0		

Individual letters make words and words make up sentences (or lines) and lines make up paragraphs. Paragraphs make up columns of text and columns of text make up pages. When handling type we are dealing with designed letterforms. Only very few experienced typographers take the time and consid-*erable effort to design new typefaces. Though over the years thousands of typefaces*

User Defined H&J

	Min	Opt	Max
W ✐	75	100	115
C ✐	0	0	10
H ✓	ON		
T ✓	2		

Individual letters make words and words make up sentences (or lines) and lines make up paragraphs. Paragraphs make up columns of text and columns of text make up pages. When handling type we are dealing with designed letterforms. Only very few experienced typographers *take the time and consider-able effort to design new type-faces. Though over the years*

User Defined H&J

	Min	Opt	Max
W ✐	85	100	250
C ✐	0	0	0
H	OFF		
T	0		

Individual letters make words and words make up sentences (or lines) and lines make up paragraphs. Paragraphs make up columns of text and columns of text make up pages. When handling type we are dealing with designed letterforms. Only very few experienced typographers take the time and considerable effort to design new typefaces. Though over the years thousands of typefaces have been designed, so the choice for us is enormous. When we look at individual typefaces we can see that they each have their own particular characteristics and "personality." We may examine each of the 26 letters of the alphabet and can see *that their form and structure has a special relationship with the background on which it sits. Not only does the shape of the letter itself give it form, but the space around it accents its*

User Defined H&J

	Min	Opt	Max
W ✐	90	100	120
C ✐	0	3	10
H	OFF		
T	0		

Individual letters make words and words make up sentences (or lines) and lines make up paragraphs. Paragraphs make up columns of text and columns of text make up pages. When handling type we are dealing with designed letterforms. Only very few experienced typographers take the time and considerable effort to design new typefaces. Though over the years thousands of typefaces have been designed, so the choice for us is enormous. When we look at individual typefaces we can see that they each have their own particular *characteristics and "personality." We may examine each of the 26 letters of the alphabet and can see that their form and structure has a special relationship with the*

ITALIC *AaBbCcDdEeFfGgHhIiJjKkLlMmNnOoPpQqRrSsTt UuVvWwXxYyZz1234567890;:'"!?@#£$&*()+=*

BOLD **AaBbCcDdEeFfGgHhIiJjKkLlMmNnOoPpQqRrSsTt UuVvWwXxYyZz1234567890;:'"!?@#£$&*()+=**

BOLD ITALIC ***AaBbCcDdEeFfGgHhIiJjKkLlMmNnOoPpQqRrSsTt UuVvWwXxYyZz1234567890;:'"!?@#£$&*()+=***

BLACK **AaBbCcDdEeFfGgHhIiJjKkLlMmNnOoPpQqRrSsTt UuVvWwXxYyZz1234567890;:'"!?@#£$&*()+=**

BLACK ITALIC ***AaBbCcDdEeFfGgHhIiJjKkLlMmNnOoPpQqRrSsTt UuVvWwXxYyZz1234567890;:'"!?@#£$&*()+=***

BOLD COND **AaBbCcDdEeFfGgHhIiJjKkLlMmNnOoPpQqRrSsTt UuVvWwXxYyZz1234567890;:'"!?@#£$&*()+=**

BLACK COND **AaBbCcDdEeFfGgHhIiJjKkLlMmNnOoPpQqRrSsTt UuVvWwXxYyZz1234567890;:'"!?@#£$&*()+=**

*qrstuvwxyz*PAGES66-67

SERIF

Bodoni

Bodoni Bold (Bauer)

Setting Tips

The Bauer version of Bodoni is considered among the closest in appearance to the original and responds to skillful handling. Bauer is comfortable at text sizes over 10 point, but text settings need to be well leaded to help balance the powerful verticality and avoid tiring reading. Bodoni is best when not closely spaced. At display sizes, it has a striking elegance. Bodoni has a small x-height, giving a greater than average number of characters to a line. It is shown here next to Clearface. The contrasting thick and thin vertical strokes allows for wider tracking to be successfully used.

SHALLOW X-HEIGHT

BB

BLACK COND

BLACK

BOLD COND

BOLD

REGULAR

BBB

AaBbCc

AaBbCcDdEeFfGgHhIiJjKkLlMmNnOoPpQqRrSsTt
UuVvWwXxYyZz1234567890;:'"!?@#£$&*()+=

Quark Default H&J

	Min	Opt	Max
W	85	110	250
C	0	0	4
H ✓	ON		
T	0		

Individual letters make words and words make up sentences (or lines) and lines make up paragraphs. Paragraphs make up columns of text and columns of text make up pages. When handling type we are dealing with designed letterforms. Only very few experienced typographers take the time and considerable effort to design new typefaces. Though over the years thousands of typefaces have been designed, so the choice for us is enormous. When we look at individual typefaces we can see that they each have their *own particular characteristics and "personality." We may examine each of the 26 letters of the alphabet and can see that their form and structure has a special rela-*

User Defined H&J

	Min	Opt	Max
W	90	100	100
C	0	0	3
H ✓	ON		
T ✓	3		

Individual letters make words and words make up sentences (or lines) and lines make up paragraphs. Paragraphs make up columns of text and columns of text make up pages. When handling type we are dealing with designed letterforms. Only very few experienced typographers take the time and considerable effort to design new typefaces. Though over the years thousands of typefaces have been designed, so the choice for us is enormous. When we look at individual typefaces we can see that they *each have their own particular characteristics and "personality." We may examine each of the 26 letters of the alphabet and can see that their form and struc-*

User Defined H&J

	Min	Opt	Max
W	90	100	120
C	0	3	10
H ✓	ON		
T	0		

User Defined H&J

	Min	Opt	Max
W	90	100	120
C	0	3	10
H	OFF		
T	0		

Individual letters make words and words make up sentences (or lines) and lines make up paragraphs. Paragraphs make up columns of text and columns of text make up pages. When handling type we are dealing with designed letterforms. Only very few experienced typographers take the time and consid-

Individual letters make words and words make up sentences (or lines) and lines make up paragraphs. Paragraphs make up columns of text and columns of text make up pages. When handling type we are dealing with designed letterforms. Only very few experienced typographers take the time and

Quark Default H&J	Min	Opt	Max
W ✐	85	110	250
C ✐	0	0	4
H ✓	ON		
T	0		

Individual letters make words and words make up sentences (or lines) and lines make up paragraphs. Paragraphs make up columns of text and columns of text make up pages. When handling type we are dealing with designed letterforms. Only *very few experienced typographers take the time and considerable effort to design*

Quark Default H&J	Min	Opt	Max
W ✐	85	110	250
C ✐	0	0	4
H	OFF		
T	0		

Individual letters make words and words make up sentences (or lines) and lines make up paragraphs. Paragraphs make up columns of text and columns of text make up pages. When handling type we are dealing with designed letterforms. Only *very few experienced typographers take the time and considerable effort to*

Quark Default H&J	Min	Opt	Max
W ✐	85	110	250
C ✐	0	0	4
H ✓	ON		
T	0		

Individual letters make words and words make up sentences (or lines) and lines make up paragraphs. Paragraphs make up columns of text and columns of text make up pages. When handling type we are dealing with designed letterforms. Only *very few experienced typographers take the time and considerable effort to design*

Quark Default H&J	Min	Opt	Max
W ✐	85	110	250
C ✐	0	0	4
H ✓	ON	HZ 14mm	
T	0		

Individual letters make words and words make up sentences (or lines) and lines make up paragraphs. Paragraphs make up columns of text and columns of text make up pages. When handling type we are dealing with designed letterforms. Only *very few experienced typographers take the time and considerable effort to*

User Defined H&J	Min	Opt	Max
W ✐	75	90	110
C ✐	–3	–3	3
H ✓	ON		
T	0		

Individual letters make words and words make up sentences (or lines) and lines make up paragraphs. Paragraphs make up columns of text and columns of text make up pages. When handling type we are dealing with designed letterforms. Only very few experienced *typographers take the time and considerable effort to design new typefaces. Though*

User Defined H&J	Min	Opt	Max
W ✐	100	100	200
C ✐	0	5	10
H ✓	ON		
T ✓	2		

Individual letters make words and words make up sentences (or lines) and lines make up paragraphs. Paragraphs make up columns of text and columns of text make up pages. When handling type we are dealing with designed letterforms. Only *very few experienced typographers take the time and considerable effort to design*

User Defined H&J	Min	Opt	Max
W ✐	85	100	250
C ✐	0	0	0
H	OFF		
T	0		

Individual letters make words and words make up sentences (or lines) and lines make up paragraphs. Paragraphs make up columns of text and columns of text make up pages. When handling type we are dealing with designed letterforms. Only very few experienced typographers take the time and considerable effort to design new typefaces. Though over the years thousands of typefaces have been designed, so the choice for us is enormous. When we look at individual typefaces we can see that they each have their own particular characteristics and "personality." *We may examine each of the 26 letters of the alphabet and can see that their form and structure has a special relationship with the background on which it sits. Not only does the shape*

Bodoni

—7 —6 —12 —9 —5

BODONI

—4 —7 —2 —9 —0

Bodoni showing overall tracking of 0 with letter pairs individually kerned as shown. Gray text indicates 0 tracking and no kerning of letter pairs.

Individual letters make words

—5 —7 —6

SERIF

Caslon 540

Caslon revives an early 18th-century English font named after its designer William Caslon, a highly skilled engraver, typefounder, and printer. Its popularity has ensured continuous use for nearly 300 years. There have been many copies and derivatives. One of the first was used to set the original Declaration of Independence and the Constitution of the United States of America.

Released at the beginning of the 20th century by ATF (American Typefounders), Caslon 540 is a lighter, graceful version than the original Caslon. With its fine feel and versatility it has an affinity with the original design. It has lots of visual interest to involve the reader, particularly in the italics, which can be used to great advantage in display sizes. It is a very good text face with an average x-height making for reasonably economical leading.

TYPE CHARACTERISTICS

This beautiful version is characterized by a measured contrast in the stroke weight, making it lighter overall with less dazzle on the page than some of the other versions. There are few common characteristics to the many versions of Caslon apart from the angled, slightly concave apex of the capital "A," the markedly full serifs on the capital "C" and the leg of the capital "R" with its minimal curve at the junction with the bowl. The tail designs of the capital "Q" vary, but the 540 version is medium length with a rhythmic flow. The stress of Caslon 540 is almost vertical.

In the lowercase letters, the design of the curved characters is a little narrow. The "e" has a high, horizontal crossbar, and the upper, thin stroke of the "k" has a very pronounced serif.

The lowercase italics have a definite angle with a lively but even pace. The "v" and "w" are round and together with the "z" and the grandiose ampersand, have a flowing, swash-like feel to them. The Caslon italic ampersand is one of great beauty and exuberance.

Small caps, Old Style numerals, italics, and swashes are available but there is no specific bold available for this version of Caslon.

Shown overleaf four versions of Caslon are compared.

ABCDEFGHIJKLMNOPQRSTUVWXYZ
abcdefghijklmnopqrstuvwxyz1234567890;:'"'!?@#£$&*()+=

Quark Default H&J

	Min	Opt	Max
W	85	110	250
C	0	0	4
H	ON		
T	0		

Individual letters make words and words make up sentences (or lines) and lines make up paragraphs. Paragraphs make up columns of text and columns of text make up pages. When handling type we are dealing with designed letterforms. Only very few experienced typographers take the time and considerable effort to design new typefaces. Though over the years thousands of typefaces have been designed, so the choice for us is enormous. When we look at individual typefaces we can see that they each have their own particular characteristics and "personality." We *may examine each of the 26 letters of the alphabet and can see that their form and structure has a special relationship with the background on which it sits. Not only does the shape of the letter itself*

User Defined H&J

	Min	Opt	Max
W	90	100	100
C	0	0	3
H	ON		
T	0		

Individual letters make words and words make up sentences (or lines) and lines make up paragraphs. Paragraphs make up columns of text and columns of text make up pages. When handling type we are dealing with designed letterforms. Only very few experienced typographers take the time and considerable effort to design new typefaces. Though over the years thousands of typefaces have been designed, so the choice for us is enormous. When we look at individual typefaces we can see that they each have their own particular characteristics and "personality." We *may examine each of the 26 letters of the alphabet and can see that their form and structure has a special relationship with the background on which it sits. Not only does the shape of the letter itself*

User Defined H&J

	Min	Opt	Max
W	75	90	90
C	−3	0	3
H	ON		
T	0		

Individual letters make words and words make up sentences (or lines) and lines make up paragraphs. Paragraphs make up columns of text and columns of text make up pages. When handling type we are dealing with designed letterforms. Only very few experienced typographers take the time and considerable effort to design new typefaces. Though over the years thousands of typefaces have been designed, so the choice for us is enormous. When we look at individual typefaces we can see that they each have their own particular characteristics and "personality." We may examine each of *the 26 letters of the alphabet and can see that their form and structure has a special relationship with the background on which it sits. Not only does the shape of the letter itself give it form, but the space*

Column 1

Quark Default H&J

	Min	Opt	Max
W	85	110	250
C	0	0	4
H ✓	ON		
T	0		

Individual letters make words and words make up sentences (or lines) and lines make up paragraphs. Paragraphs make up columns of text and columns of text make up pages. When handling type we are dealing with designed letterforms. Only very few experienced typographers take *the time and considerable effort to design new typefaces. Though over the years thousands of typefaces*

User Defined H&J

	Min	Opt	Max
W	75	90	110
C	-3	-3	3
H ✓	ON		
T	0		

Individual letters make words and words make up sentences (or lines) and lines make up paragraphs. Paragraphs make up columns of text and columns of text make up pages. When handling type we are dealing with designed letterforms. Only very few experienced typographers take the time and consid-*erable effort to design new typefaces. Though over the years thousands of typefaces have been designed, so the*

User Defined H&J

	Min	Opt	Max
W	90	100	120
C	0	3	10
H	OFF		
T	0		

Individual letters make words and words make up sentences (or lines) and lines make up paragraphs. Paragraphs make up columns of text and columns of text make up pages. When handling type we are dealing with designed letterforms. Only very few experienced typographers take the time and considerable effort to design new typefaces. Though over the years thousands of typefaces have been designed, so the choice for us is enormous. When we look at individual typefaces we can see that they each have their own particular characteristics *and "personality." We may examine each of the 26 letters of the alphabet and can see that their form and structure has a special relationship with the background on which it sits. Not only does*

Column 2

Quark Default H&J

	Min	Opt	Max
W	85	110	250
C	0	0	4
H	OFF		
T	0		

Individual letters make words and words make up sentences (or lines) and lines make up paragraphs. Paragraphs make up columns of text and columns of text make up pages. When handling type we are dealing with designed letterforms. Only very few experienced typographers take *the time and considerable effort to design new typefaces. Though over the years thousands of typefaces*

User Defined H&J

	Min	Opt	Max
W	75	100	115
C	0	0	10
H ✓	ON		
T ✓	2		

Individual letters make words and words make up sentences (or lines) and lines make up paragraphs. Paragraphs make up columns of text and columns of text make up pages. When handling type we are dealing with designed letterforms. Only very few experienced typographers *take the time and considerable effort to design new typefaces. Though over the years thousands*

Column 3

Quark Default H&J

	Min	Opt	Max
W	85	110	250
C	0	0	4
H ✓	ON		
T	0		

Individual letters make words and words make up sentences (or lines) and lines make up paragraphs. Paragraphs make up columns of text and columns of text make up pages. When handling type we are dealing with designed letterforms. Only very few experienced typographers *take the time and considerable effort to design new typefaces. Though over the years thousands*

User Defined H&J

	Min	Opt	Max
W	85	100	250
C	0	0	0
H	OFF		
T	0		

Individual letters make words and words make up sentences (or lines) and lines make up paragraphs. Paragraphs make up columns of text and columns of text make up pages. When handling type we are dealing with designed letterforms. Only very few experienced typographers take the time and considerable effort to design new typefaces. Though over the years thousands of typefaces have been designed, so the choice for us is enormous. When we look at individual typefaces we can see that they each have their own particular characteristics and "personality." We may examine each of the 26 letters of the alphabet and can see *that their form and structure has a special relationship with the background on which it sits. Not only does the shape of the letter itself give it form, but the space around it accents its individuality. When two or*

Column 4

Quark Default H&J

	Min	Opt	Max
W	85	110	250
C	0	0	4
H ✓	ON	HZ 14mm	
T	0		

Individual letters make words and words make up sentences (or lines) and lines make up paragraphs. Paragraphs make up columns of text and columns of text make up pages. When handling type we are dealing with designed letterforms. Only very few experienced typographers *take the time and considerable effort to design new typefaces. Though over the years thousands of typefaces*

Type specimens

ITALIC *AaBbCcDdEeFfGgHhIiJjKkLlMmNnOoPpQqRrSsTt UuVvWwXxYyZz1234567890;:'"!?@#£$&*()+=*

SMALL CAPS AaBbCcDdEeFfGgHhIiJjKkLlMmNnOoPpQqRrSsTt UuVvWwXxYyZz1234567890;:'"!?@#£$&*()+=

ITALIC, OLD STYLE FIGURES *AaBbCcDdEeFfGgHhIiJjKkLlMmNnOoPpQqRrSsTt UuVvWwXxYyZz1234567890;:'"!?@#£$&*()+=*

Caslon revivals

Caslon

Setting Tips

Caslon is particularly readable in blocks of running text set to a generous measure. Variation in tracking should be modest to maintain Caslon's rhythm and readability. Adobe Caslon and Caslon 224 do have bold versions, and Caslon 3 can be used as a bold companion to Caslon 540 if care is taken to match the x-heights.

Four versions of Caslon showing a wide variety of letterforms, particularly the capital "Q"s. Adobe Caslon (shown on the left) has some fine ligatures such as "ct," "st," and "ft."

AaBbCc
ABCDEFGHIJKLMNOPQRSTUVWXYZ
abcdefghijklmnopqrstuvwxyz1234567890;:'"!?@#£$&*()+=

CASLON 540

ABCDEFGHIJKLMNOPQRSTUVWXYZ
abcdefghijklmnopqrstuvwxyz
1234567890;:'"!?@#£$&*()+=
ABCDEFGHIJKLMNOPQRSTUVWXYZ
abcdefghijklmnopqrstuvwxyz
1234567890;:'"!?@#£$&()+=*

ADOBE CASLON

ABCDEFGHIJKLMNOPQRSTUVWXYZ
abcdefghijklmnopqrstuvwxyz
1234567890;:'"!?@#£$&*()+=
ABCDEFGHIJKLMNOPQRSTUVWXYZ
abcdefghijklmnopqrstuvwxyz
1234567890;:'"!?@#£$&()+=*

CASLON 224

ABCDEFGHIJKLMNOPQRSTUVWXYZ
abcdefghijklmnopqrstuvwxyz
1234567890;:'"!?@#£$&*()+=
ABCDEFGHIJKLMNOPQRSTUVWXYZ
abcdefghijklmnopqrstuvwxyz
1234567890;:'"!?@#£$&()+=*

CASLON 3

ABCDEFGHIJKLMNOPQRSTUVWXYZ
abcdefghijklmnopqrstuvwxyz
1234567890;:'"!?@#£$&*()+=
ABCDEFGHIJKLMNOPQRSTUVWXYZ
abcdefghijklmnopqrstuvwxyz
1234567890;:'"!?@#£$&()+=*

Caslon 540

Individual letters make words and words make up sentences (or lines) and lines make up paragraphs. Paragraphs make up columns of text and columns of text make up pages. When handling type we are dealing with designed letterforms. Only very few experienced typographers take the time and considerable effort to design new typefaces. Though over the years thousands of typefaces have been

Caslon ADOBE

Individual letters make words and words make up sentences (or lines) and lines make up paragraphs. Paragraphs make up columns of text and columns of text make up pages. When handling type we are dealing with designed letterforms. Only very few experienced typographers take the time and considerable effort to design new typefaces. Though over the years thousands of typefaces have been designed, so the choice for us is

Caslon 224

Individual letters make words and words make up sentences (or lines) and lines make up paragraphs. Paragraphs make up columns of text and columns of text make up pages. When handling type we are dealing with designed letterforms. Only very few experienced typographers take the time and considerable effort to design new typefaces. Though over the years thousands of typefaces have been

Caslon 3

Individual letters make words and words make up sentences (or lines) and lines make up paragraphs. Paragraphs make up columns of text and columns of text make up pages. When handling type we are dealing with designed letterforms. Only very few experienced typographers take the time and considerable effort to design new typefaces. Though over the years

ITC Clearface

The original Clearface, on which ITC Clearface is based, was produced by the company American Type Founders in 1907. Designed primarily for use in display contexts by a major figure in American type design, Morris Fuller Benton, Clearface is the functional, easy-to-read typeface its name suggests.

In 1978, Victor Caruso was commissioned by the International Typeface Corporation to extend and develop Benton's original utilitarian Clearface design into a more modern family of four weights: regular, bold, heavy, and black, with corresponding italics. A key aspect of Caruso's ITC Clearface design was a focus on text setting, further extending its use.

TYPE CHARACTERISTICS

Its distinctive appearance projects a warm, friendly mood, contributing to its continuing popularity. ITC Clearface retains the characteristic sloping bar on the lowercase "e," its axis inclining to the left, and the small contrast in the thick and thin strokes of the letters. Although slightly condensed, it has a large x-height, small wedge-like serifs, and short descenders, which make it a very legible typeface and an economical choice when space is at a premium. It is also useful for narrow column settings. Some extra tracking may be required in the bold weight at smaller sizes.

Recognizable characters include the capital "C" and "G" with high arches and left-inclining stress. The capital "V" and "W" have over-running thick strokes, which gives them the appearance of having small foot serifs. In the lowercase letters, the stems have oblique serifs. The top of the arm of the "a" inclines upward and the "r" has an angled stem. The terminals (end of the strokes) of the "a," "c," "k," "s," "v," "w," and "y" are rounded.

ABCDEFGHIJKLMNOPQRSTUVWXYZ
abcdefghijklmnopqrstuvwxyz1234567890;:'"!?@#£$&*()+=

Quark Default H&J

	Min	Opt	Max
W ✎	85	110	250
C ✎	0	0	4
H ✓	ON		
T	0		

Individual letters make words and words make up sentences (or lines) and lines make up paragraphs. Paragraphs make up columns of text and columns of text make up pages. When handling type we are dealing with designed letterforms. Only very few experienced typographers take the time and considerable effort to design new typefaces. Though over the years thousands of typefaces have been designed, so the choice for us is enormous. When we look at individual typefaces we can see that they each have their own particular characteristics and "personality." We may examine each of *the 26 letters of the alphabet and can see that their form and structure has a special relationship with the background on which it sits. Not only does the shape of the letter*

User Defined H&J

	Min	Opt	Max
W ✎	75	100	105
C ✎	0	0	0
H ✓	ON		
T ✓	3		

Individual letters make words and words make up sentences (or lines) and lines make up paragraphs. Paragraphs make up columns of text and columns of text make up pages. When handling type we are dealing with designed letterforms. Only very few experienced typographers take the time and considerable effort to design new typefaces. Though over the years thousands of typefaces have been designed, so the choice for us is enormous. When we look at individual typefaces we can see that they each have their own particular characteristics and "personality." We may *examine each of the 26 letters of the alphabet and can see that their form and structure has a special relationship with the background on which it sits. Not only does the*

User Defined H&J

	Min	Opt	Max
W ✎	90	100	120
C ✎	0	3	10
H ✓	ON		
T	0		

Individual letters make words and words make up sentences (or lines) and lines make up paragraphs. Paragraphs make up columns of text and columns of text make up pages. When handling type we are dealing with designed letterforms. Only very few experienced typographers take the time and considerable effort to design new typefaces. Though over the years thousands of typefaces have been designed, so the choice for us is enormous. When we look at individual typefaces we can see that they each have their own particular characteristics and "personality." We may *examine each of the 26 letters of the alphabet and can see that their form and structure has a special relationship with the background on which it sits. Not only does the*

Quark Default H&J

	Min	Opt	Max
W	85	110	250
C	0	0	4
H ✓	ON		
T	0		

Individual letters make words and words make up sentences (or lines) and lines make up paragraphs. Paragraphs make up columns of text and columns of text make up pages. When handling type we are dealing with designed letterforms. Only very few experienced typographers take the *time and considerable effort to design new typefaces. Though over the years thousands of*

Quark Default H&J

	Min	Opt	Max
W	85	110	250
C	0	0	4
H	OFF		
T	0		

Individual letters make words and words make up sentences (or lines) and lines make up paragraphs. Paragraphs make up columns of text and columns of text make up pages. When handling type we are dealing with designed letterforms. Only very few experienced typographers take *the time and considerable effort to design new typefaces. Though over the years*

Quark Default H&J

	Min	Opt	Max
W	85	110	250
C	0	0	4
H ✓	ON		
T	0		

Individual letters make words and words make up sentences (or lines) and lines make up paragraphs. Paragraphs make up columns of text and columns of text make up pages. When handling type we are dealing with designed letterforms. Only very few experienced typographers take *the time and considerable effort to design new typefaces. Though over the years thou-*

Quark Default H&J

	Min	Opt	Max
W	85	110	250
C	0	0	4
H ✓	ON	HZ 14mm	
T	0		

Individual letters make words and words make up sentences (or lines) and lines make up paragraphs. Paragraphs make up columns of text and columns of text make up pages. When handling type we are dealing with designed letterforms. Only very few experienced typographers take *the time and considerable effort to design new typefaces. Though over the years*

User Defined H&J

	Min	Opt	Max
W	75	100	115
C	0	0	10
H ✓	ON		
T	0		

Individual letters make words and words make up sentences (or lines) and lines make up paragraphs. Paragraphs make up columns of text and columns of text make up pages. When handling type we are dealing with designed letterforms. Only very few experienced typogra-phers take the time and consid-*erable effort to design new typefaces. Though over the years thousands of typefaces*

User Defined H&J

	Min	Opt	Max
W	85	100	120
C	–3	0	–3
H ✓	ON		
T ✓	2		

Individual letters make words and words make up sentences (or lines) and lines make up paragraphs. Paragraphs make up columns of text and columns of text make up pages. When handling type we are dealing with designed letterforms. Only very few experienced typogra-phers take the time and consid-*erable effort to design new typefaces. Though over the years thousands of* typefaces

User Defined H&J

	Min	Opt	Max
W	85	100	250
C	0	0	0
H	OFF		
T	0		

Individual letters make words and words make up sentences (or lines) and lines make up paragraphs. Paragraphs make up columns of text and columns of text make up pages. When handling type we are dealing with designed letterforms. Only very few experienced typographers take the time and considerable effort to design new typefaces. Though over the years thousands of typefaces have been designed, so the choice for us is enormous. When we look at individual typefaces we can see that they each have their own particular characteristics and "personality." We may examine each of the 26 letters of the alphabet and can see that their form and *structure has a special relationship with the background on which it sits. Not only does the shape of the letter itself give it form, but the space around it accents its individuality. When two or more*

User Defined H&J

	Min	Opt	Max
W	75	100	105
C	0	0	0
H	OFF		
T	0		

Individual letters make words and words make up sentences (or lines) and lines make up paragraphs. Paragraphs make up columns of text and columns of text make up pages. When handling type we are dealing with designed letterforms. Only very few experienced typographers take the time and considerable effort to design new typefaces. Though over the years thousands of typefaces have been designed, so the choice for us is enormous. When we look at individual typefaces we can see that they each have their own particular characteristics and "personality." We may examine each of the *26 letters of the alphabet and can see that their form and structure has a special relationship with the background on which it sits. Not only does the shape of the letter itself give*

ITALIC *AaBbCcDdEeFfGgHhIiJjKkLlMmNnOoPpQqRrSsTt UuVvWwXxYyZz1234567890;:'"!?@#£$&*()+=*

BOLD **AaBbCcDdEeFfGgHhIiJjKkLlMmNnOoPpQqRrSsTt UuVvWwXxYyZz1234567890;:'"!?@#£$&*()+=**

BOLD ITALIC ***AaBbCcDdEeFfGgHhIiJjKkLlMmNnOoPpQqRrSsTt UuVvWwXxYyZz1234567890;:'"!?@#£$&*()+=***

HEAVY **AaBbCcDdEeFfGgHhIiJjKkLlMmNnOoPpQqRrSsTt UuVvWwXxYyZz1234567890;:'"!?@#£$&*()+=**

HEAVY ITALIC ***AaBbCcDdEeFfGgHhIiJjKkLlMmNnOoPpQqRrSsTt UuVvWwXxYyZz1234567890;:'"!?@#£$&*()+=***

BLACK **AaBbCcDdEeFfGgHhIiJjKkLlMmNnOoPpQqRrSsTt UuVvWwXxYyZz1234567890;:'"!?@#£$&*()+=**

BLACK ITALIC ***AaBbCcDdEeFfGgHhIiJjKkLlMmNnOoPpQqRrSsTt UuVvWwXxYyZz1234567890;:'"!?@#£$&*()+=***

SERIF

Ehrhardt

Ehrhardt, issued in 1938 by Monotype, was one of the last major Monotype revivals. It came into existence as a result of Monotype's desire to recut Janson—an early 18th-century design thought to be by Anton Janson, but later discovered to have been cut by Nicholas Kis, a Hungarian living in the Netherlands. Stanley Morison, Monotype's typographic advisor, suggested a larger and slightly narrower form. The outcome was Ehrhardt, named after the German foundry in Leipzig where the Janson types were once kept.

Ehrhardt has a less marked change in stroke weight and a greater regularity than Janson. The large x-height, shorter ascenders and descenders, combined with its clean, clear form, make it an excellent and highly readable text face particularly suited to book texts. Its minimally condensed character also makes it a good, legible choice when space is at a premium.

TYPE CHARACTERISTICS

Ehrhardt has a typical crispness of cut to its letters. Recognizable characteristics in the capital letters include, a slightly angled apex and distinctively bowing crossbar to the "A," a similar bowing to the middle junction of the "B." The stems of the capital "M" are splayed, and have a minimally bracketed serif, the capital "P" has an angled junction, and the crossed centerstrokes of the capital "W" join just below the cap height.

In the lowercase letters, the arm of the "a" is angled, the base of the "b" rounded, the loop of the "g" is angled backward, and the "j" has a short tapered tail.

In the italics, the serifs are elegantly angled, and the terminals of the lowercase "w" are curved backward giving it an exaggerated italic angle. The lowercase "v" and "z" have some similarity to those of Caslon 540.

The Ehrhardt family includes a regular and a semibold, both with italic versions, and an Expert Set.

ABCDEFGHIJKLMNOPQRSTUVWXYZ
abcdefghijklmnopqrstuvwxyz1234567890;:'"'!?@#£$&*()+=

Quark Default H&J

	Min	Opt	Max
W ✎	85	110	250
C ✎	0	0	4
H ✓	ON		
T	0		

Individual letters make words and words make up sentences (or lines) and lines make up paragraphs. Paragraphs make up columns of text and columns of text make up pages. When handling type we are dealing with designed letterforms. Only very few experienced typographers take the time and considerable effort to design new typefaces. Though over the years thousands of typefaces have been designed, so the choice for us is enormous. When we look at individual typefaces we can see that they each have their own particular characteristics and "personality." We may examine each of the 26 letters of the *alphabet and can see that their form and structure has a special relationship with the background on which it sits. Not only does the shape of the letter itself give it form, but the space around it accents*

User Defined H&J

	Min	Opt	Max
W ✎	90	100	105
C ✎	0	0	3
H ✓	ON		
T ✓	3		

Individual letters make words and words make up sentences (or lines) and lines make up paragraphs. Paragraphs make up columns of text and columns of text make up pages. When handling type we are dealing with designed letterforms. Only very few experienced typographers take the time and considerable effort to design new typefaces. Though over the years thousands of typefaces have been designed, so the choice for us is enormous. When we look at individual typefaces we can see that they each have their own particular characteristics and "personality." We may examine each of the 26 letters of *the alphabet and can see that their form and structure has a special relationship with the background on which it sits. Not only does the shape of the letter itself give it form, but the space*

User Defined H&J

	Min	Opt	Max
W ✎	.75	100	105
C ✎	0	0	0
H ✓	ON		
T	0		

Individual letters make words and words make up sentences (or lines) and lines make up paragraphs. Paragraphs make up columns of text and columns of text make up pages. When handling type we are dealing with designed letterforms. Only very few experienced typographers take the time and considerable effort to design new typefaces. Though over the years thousands of typefaces have been designed, so the choice for us is enormous. When we look at individual typefaces we can see that they each have their own particular characteristics and "personality." We may examine each of the 26 letters of the alphabet and can *see that their form and structure has a special relationship with the background on which it sits. Not only does the shape of the letter itself give it form, but the space around it accents its individuality*

Quark Default H&J	Min	Opt	Max
W ✎	85	110	250
C ✎	0	0	4
H ✓	ON		
T	0		

Individual letters make words and words make up sentences (or lines) and lines make up paragraphs. Paragraphs make up columns of text and columns of text make up pages. When handling type we are dealing with designed letterforms. Only very few experienced typographers take the time and considerable *effort to design new typefaces. Though over the years thousands of typefaces have been designed, so the*

Quark Default H&J	Min	Opt	Max
W ✎	85	110	250
C ✎	0	0	4
H	OFF		
T	0		

Individual letters make words and words make up sentences (or lines) and lines make up paragraphs. Paragraphs make up columns of text and columns of text make up pages. When handling type we are dealing with designed letterforms. Only very few experienced typographers take the time and *considerable effort to design new typefaces. Though over the years thousands of typefaces have been*

Quark Default H&J	Min	Opt	Max
W ✎	85	110	250
C ✎	0	0	4
H ✓	ON		
T	0		

Individual letters make words and words make up sentences (or lines) and lines make up paragraphs. Paragraphs make up columns of text and columns of text make up pages. When handling type we are dealing with designed letterforms. Only very few experienced typographers take the time and considerable *effort to design new typefaces. Though over the years thousands of typefaces have been designed,* so

Quark Default H&J	Min	Opt	Max
W ✎	85	110	250
C ✎	0	0	4
H ✓	ON HZ 14mm		
T	0		

Individual letters make words and words make up sentences (or lines) and lines make up paragraphs. Paragraphs make up columns of text and columns of text make up pages. When handling type we are dealing with designed letterforms. Only very few experienced typographers take the time and *considerable effort to design new typefaces. Though over the years thousands of typefaces have been*

User Defined H&J	Min	Opt	Max
W ✎	75	90	110
C ✎	–3	–3	3
H ✓	ON		
T	0		

Individual letters make words and words make up sentences (or lines) and lines make up paragraphs. Paragraphs make up columns of text and columns of text make up pages. When handling type we are dealing with designed letterforms. Only very few experienced typographers take the time and considerable effort to design new *typefaces. Though over the years thousands of typefaces have been designed, so the choice for us is enor-*

User Defined H&J	Min	Opt	Max
W ✎	75	100	115
C ✎	0	0	10
H ✓	ON		
T ✓	2		

Individual letters make words and words make up sentences (or lines) and lines make up paragraphs. Paragraphs make up columns of text and columns of text make up pages. When handling type we are dealing with designed letterforms. Only very few experienced typographers take the time and considerable *effort to design new typefaces. Though over the years thousands of typefaces have been designed, so*

User Defined H&J	Min	Opt	Max
W ✎	85	100	250
C ✎	0	0	0
H	OFF		
T	0		

Individual letters make words and words make up sentences (or lines) and lines make up paragraphs. Paragraphs make up columns of text and columns of text make up pages. When handling type we are dealing with designed letterforms. Only very few experienced typographers take the time and considerable effort to design new typefaces. Though over the years thousands of typefaces have been designed, so the choice for us is enormous. When we look at individual typefaces we can see that they each have their own particular characteristics and "personality." We may examine each of the 26 letters of the alphabet and can see that their form and structure has a special relationship with *the background on which it sits. Not only does the shape of the letter itself give it form, but the space around it accents its individuality. When two or more letters are put side by side they create words which are recognised as*

User Defined H&J	Min	Opt	Max
W ✎	80	105	150
C ✎	0	3	3
H	OFF		
T	0		

Individual letters make words and words make up sentences (or lines) and lines make up paragraphs. Paragraphs make up columns of text and columns of text make up pages. When handling type we are dealing with designed letterforms. Only very few experienced typographers take the time and considerable effort to design new typefaces. Though over the years thousands of typefaces have been designed, so the choice for us is enormous. When we look at individual typefaces we can see that they each have their own particular characteristics and "personality." We may examine each of *the 26 letters of the alphabet and can see that their form and structure has a special relationship with the background on which it sits. Not only does the shape of the letter itself give it form, but*

ITALIC *AaBbCcDdEeFfGgHhIiJjKkLlMmNnOoPpQqRrSsTt UuVvWwXx YyZz1234567890;:'"!?@#£$&*()+=*

SEMIBOLD **AaBbCcDdEeFfGgHhIiJjKkLlMmNnOoPpQqRrSsTt UuVvWwXxYyZz1234567890;:'"!?@#£$&*()+=**

SEMIBOLD ITALIC ***AaBbCcDdEeFfGgHhIiJjKkLlMmNnOoPpQqRrSsTt UuVvWwXx YyZz1234567890;:'"!?@#£$&*()+=***

EXPERT REGULAR A B CĐD E F¼G½H¾I⅛J⅜K⅝L⅞M⅓N⅔O P QʳR S T uffvfiwflxffiyfflz1234567890;: !? ¢₃$&..⁰.—

EXPERT ITALIC *¼ ½ ¾ ⅛ ⅜ ⅝ ⅞ ⅓ ⅔ ff fi fl ffi ffl 1234567890;: ¢₃$ ..⁰.—*

EXPERT SEMIBOLD **¼ ½ ¾ ⅛ ⅜ ⅝ ⅞ ⅓ ⅔ ff fi fl ffi ffl 1234567890;: ¢₃$ ..⁰.—**

EXPERT SEMIBOLD ITALIC ***¼ ½ ¾ ⅛ ⅜ ⅝ ⅞ ⅓ ⅔ ff fi fl ffi ffl 1234567890;: ¢₃$ ..⁰.—***

Enigma

Enigma, designed by Jeremy Tankard in 1999, is a typeface that introduces contrasting subtleties concealed within the detailing of individual forms, lending a quirkiness to the design.

Normally the design of an alphabet of capital seriffed letters includes serifs on every letter, by contrast to the lowercase design, which does not. Tankard has started to challenge this convention by including some capital letters with serif-less detailing.

The sharp "cuts" of some of the curved lowercase letterforms show an influence of 15th-century Rotunda type, a type of black letter. This unexpected detailing helps to give a distinct, slightly sparkly texture to text settings but also contributes to the linear rhythm.

Tankard has designed this interesting typeface with the range of screen display quality very much in mind, particularly the appearance of text sizes on word processors.

TYPE CHARACTERISTICS

The capitals, with their relatively narrow design and height that falls short of the ascender height, are comparatively unimposing and so sit comfortably in blocks of text.

Virtually every letterform in this typeface has a unique characteristic. Generally it is the mix of seriffed and nonseriffed strokes contained in one letterform that is striking. The lowercase "a" has a flat stroke to the top of its bowl and the crossbars of the lowercase "t" and "f" are tapered on the left.

The Enigma family includes regular and bold weights, each with small caps and italic versions, and includes all the standard ISO/Adobe characters. Refreshingly, nonlining numerals are included in the standard font, with the lining version included in the small caps font. Range availability is dependent on the type of software/hardware platform.

ABCDEFGHIJKLMNOPQRSTUVWXYZ
abcdefghijklmnopqrstuvwxyz1234567890;:'"!?@#£$&*()+=
ABCDEFGHIJKLMNOPQRSTUVWXYZ1234567890

Quark Default H&J

	Min	Opt	Max
W	85	110	250
C	0	0	4
H	ON		
T	0		

Individual letters make words and words make up sentences (or lines) and lines make up paragraphs. Paragraphs make up columns of text and columns of text make up pages. When handling type we are dealing with designed letterforms. Only very few experienced typographers take the time and considerable effort to design new typefaces. Though over the years thousands of typefaces have been designed, so the choice for us is enormous. When we look at individual typefaces we can see that they each have their own particular characteristics and "personality." We may *examine each of the 26 letters of the alphabet and can see that their form and structure has a special relationship with the background on which it sits. Not only does the shape of the letter itself give it form,*

User Defined H&J

	Min	Opt	Max
W	70	100	125
C	0	0	2
H	ON		
T	3		

Individual letters make words and words make up sentences (or lines) and lines make up paragraphs. Paragraphs make up columns of text and columns of text make up pages. When handling type we are dealing with designed letterforms. Only very few experienced typographers take the time and considerable effort to design new typefaces. Though over the years thousands of typefaces have been designed, so the choice for us is enormous. When we look at individual typefaces we can see that they each have their own particular characteristics and "personality." We may *examine each of the 26 letters of the alphabet and can see that their form and structure has a special relationship with the background on which it sits. Not only does the shape of the letter itself give it*

User Defined H&J

	Min	Opt	Max
W	80	105	150
C	0	3	3
H	ON		
T	0		

Individual letters make words and words make up sentences (or lines) and lines make up paragraphs. Paragraphs make up columns of text and columns of text make up pages. When handling type we are dealing with designed letterforms. Only very few experienced typographers take the time and considerable effort to design new typefaces. Though over the years thousands of typefaces have been designed, so the choice for us is enormous. When we look at individual typefaces we can see that they each have their own particular characteristics and "personality." We may *examine each of the 26 letters of the alphabet and can see that their form and structure has a special relationship with the background on which it sits. Not only does the shape of the letter itself give it*

Quark Default H&J

	Min	Opt	Max
W	85	110	250
C	0	0	4
H	✓ ON		
T	0		

Individual letters make words and words make up sentences (or lines) and lines make up paragraphs. Paragraphs make up columns of text and columns of text make up pages. When handling type we are dealing with designed letterforms. Only very few experienced typographers take *the time and considerable effort to design new typefaces. Though over the years thousands of typefaces*

Quark Default H&J

	Min	Opt	Max
W	85	110	250
C	0	0	4
H	OFF		
T	0		

Individual letters make words and words make up sentences (or lines) and lines make up paragraphs. Paragraphs make up columns of text and columns of text make up pages. When handling type we are dealing with designed letterforms. Only very few experienced typographers take *the time and considerable effort to design new typefaces. Though over the years thousands of typefaces*

Quark Default H&J

	Min	Opt	Max
W	85	110	250
C	0	0	4
H	✓ ON		
T	0		

Individual letters make words and words make up sentences (or lines) and lines make up paragraphs. Paragraphs make up columns of text and columns of text make up pages. When handling type we are dealing with designed letterforms. Only very few experienced typographers take *the time and considerable effort to design new typefaces. Though over the years thousands of typefaces*

Quark Default H&J

	Min	Opt	Max
W	85	110	250
C	0	0	4
H	✓ ON HZ 14mm		
T	0		

Individual letters make words and words make up sentences (or lines) and lines make up paragraphs. Paragraphs make up columns of text and columns of text make up pages. When handling type we are dealing with designed letterforms. Only very few experienced typographers take *the time and considerable effort to design new typefaces. Though over the years thousands of typefaces*

User Defined H&J

	Min	Opt	Max
W	75	90	90
C	–3	0	3
H	✓ ON		
T	0		

Individual letters make words and words make up sentences (or lines) and lines make up paragraphs. Paragraphs make up columns of text and columns of text make up pages. When handling type we are dealing with designed letterforms. Only very few experienced typographers take the time and considerable *effort to design new typefaces. Though over the years thousands of typefaces have been designed, so the*

User Defined H&J

	Min	Opt	Max
W	85	100	120
C	–3	0	3
H	✓ ON		
T	✓ 2		

Individual letters make words and words make up sentences (or lines) and lines make up paragraphs. Paragraphs make up columns of text and columns of text make up pages. When handling type we are dealing with designed letterforms. Only very few experienced typographers take the *time and considerable effort to design new typefaces. Though over the years thousands of typefaces*

User Defined H&J

	Min	Opt	Max
W	85	100	250
C	0	0	0
H	OFF		
T	0		

Individual letters make words and words make up sentences (or lines) and lines make up paragraphs. Paragraphs make up columns of text and columns of text make up pages. When handling type we are dealing with designed letterforms. Only very few experienced typographers take the time and considerable effort to design new typefaces. Though over the years thousands of typefaces have been designed, so the choice for us is enormous. When we look at individual typefaces we can see that they each have their own particular characteristics and "personality." We may examine each of the 26 letters of the alphabet and can see that their form and *structure has a special relationship with the background on which it sits. Not only does the shape of the letter itself give it form, but the space around it accents its individuality. When two or more letters are put side by side they*

User Defined H&J

	Min	Opt	Max
W	70	100	125
C	0	0	2
H	OFF		
T	0		

Individual letters make words and words make up sentences (or lines) and lines make up paragraphs. Paragraphs make up columns of text and columns of text make up pages. When handling type we are dealing with designed letterforms. Only very few experienced typographers take the time and considerable effort to design new typefaces. Though over the years thousands of typefaces have been designed, so the choice for us is enormous. When we look at individual typefaces we can see that they each have their own particular characteristics and "personality." We may *examine each of the 26 letters of the alphabet and can see that their form and structure has a special relationship with the background on which it sits. Not only does the shape of the letter itself give it form,*

ITALIC
AaBbCcDdEeFfGgHhIiJjKkLlMmNnOoPpQqRrSsTt UuVvWwXxYyZz1234567890;:'"!?@#£$&()+=*

ITALIC SMALL CAPS
AaBbCcDdEeFfGgHhIiJjKkLlMmNnOoPpQqRrSsTt UuVvWwXxYyZz1234567890;:'"!?@#£$&()+=*

BOLD
AaBbCcDdEeFfGgHhIiJjKkLlMmNnOoPpQqRrSsTt UuVvWwXxYyZz1234567890;:'"!?@#£$&*()+=

BOLD SMALL CAPS
AaBbCcDdEeFfGgHhIiJjKkLlMmNnOoPpQqRrSsTt UuVvWwXxYyZz1234567890;:'"!?@#£$&*()+=

BOLD ITALIC
AaBbCcDdEeFfGgHhIiJjKkLlMmNnOoPpQqRrSsTt UuVvWwXxYyZz1234567890;:'"!?@#£$&*()+=

BOLD ITALIC SMALL CAPS
AaBbCcDdEeFfGgHhIiJjKkLlMmNnOoPpQqRrSsTt UuVvWwXxYyZz1234567890;:'"!?@#£$&*()+=

SERIF

Monotype Garamond

The Garamond revival, named after the French Renaissance type designer Claude Garamond, was thought to follow the collection of types at the Imprimerie Nationale in Paris, known as the *caractères de l'Université*. However, after the issue of Monotype Garamond in 1922, it was discovered that the model used was a copy cut by a relatively unknown French printer, Jean Jannon, some sixty years after Garamond's death.

The adaptation of Jannon's roman "Garamond" into a version compatible with the constraints of the Monotype machinery is true to the "original." However, the distinctive italics are separately based on the work of the French punch cutter, Robert Granjon, and include a wonderful ampersand.

Most skillful manipulation of the character widths was needed to fit the grid construction of the diecase without detracting from the esthetic. Fritz Steltzer, an expert draftsman brought over from Germany, meticulously worked the Jannon copy and Granjon italics into Monotype Garamond as we know it.

TYPE CHARACTERISTICS

Almost all of the Garamond versions, including the Monotype version, have wide serifs that are slightly concave, particularly in the capital letters, which have a subtle quirkiness. In keeping with the Old Style or Garalde group of typefaces to which it belongs, Monotype Garamond has contrast between the thick and thin stroke weight, a high horizontal bar to the lowercase "e" and an oblique finish to the ascenders. It has a good visual flow making it a useful choice for text settings, and makes for effortless reading.

Among the roman capitals the "K" is wide with a single junction. The capital "T" and "Z" have the left-hand top serif at an oblique angle and the capital "R" has a narrow bowl and a long leg. The capital "Q" has a medium-length tail that sits outside the bowl.

In the roman lowercase letters, the "a" and "e" have unusually small counters, and the "j" has a medium-length, rather straight tail. The foot serifs of the lowercase "b" and "d" are angled.

ABCDEFGHIJKLMNOPQRSTUVWXYZ
abcdefghijklmnopqrstuvwxyz1234567890;:'"'!?@#£$&*()+=

Individual letters make words and words make up sentences (or lines) and lines make up paragraphs. Paragraphs make up columns of text and columns of text make up pages. When handling type we are dealing with designed letterforms. Only very few experienced typographers take the time and considerable effort to design new typefaces. Though over the years thousands of typefaces have been designed, so the choice for us is enormous. When we look at individual typefaces we can see that they each have their own particular characteristics and "personality." We may examine each of the 26 letters of the alphabet and can *see that their form and structure has a special relationship with the background on which it sits. Not only does the shape of the letter itself give it form, but the space around it accents its individuality. When two or more*

Individual letters make words and words make up sentences (or lines) and lines make up paragraphs. Paragraphs make up columns of text and columns of text make up pages. When handling type we are dealing with designed letterforms. Only very few experienced typographers take the time and considerable effort to design new typefaces. Though over the years thousands of typefaces have been designed, so the choice for us is enormous. When we look at individual typefaces we can see that they each have their own particular characteristics and "personality." We may examine each of the 26 letters of the *alphabet and can see that their form and structure has a special relationship with the background on which it sits. Not only does the shape of the letter itself give it form, but the space around it accents its individuality.*

Individual letters make words and words make up sentences (or lines) and lines make up paragraphs. Paragraphs make up columns of text and columns of text make up pages. When handling type we are dealing with designed letterforms. Only very few experienced typographers take the time and considerable effort to design new typefaces. Though over the years thousands of typefaces have been designed, so the choice for us is enormous. When we look at individual typefaces we can see that they each have their own particular characteristics and "personality." We may examine each of the 26 letters of the *alphabet and can see that their form and structure has a special relationship with the background on which it sits. Not only does the shape of the letter itself give it form, but the space around it accents its indi-*

	Min	Opt	Max
W ✐	85	110	250
C ✐	0	0	4
H ✓	ON		
T	0		

Individual letters make words and words make up sentences (or lines) and lines make up paragraphs. Paragraphs make up columns of text and columns of text make up pages. When handling type we are dealing with designed letterforms. Only very few experienced typographers take the time and considerable *effort to design new typefaces. Though over the years thousands of typefaces have been designed, so the choice for us is*

Quark Default H&J

	Min	Opt	Max
W ✐	85	110	250
C ✐	0	0	4
H	OFF		
T	0		

Individual letters make words and words make up sentences (or lines) and lines make up paragraphs. Paragraphs make up columns of text and columns of text make up pages. When handling type we are dealing with designed letterforms. Only very few experienced typographers take the time and considerable *effort to design new typefaces. Though over the years thousands of typefaces have been designed, so the choice for us is*

Quark Default H&J

	Min	Opt	Max
W ✐	85	110	250
C ✐	0	0	4
H ✓	ON		
T	0		

Individual letters make words and words make up sentences (or lines) and lines make up paragraphs. Paragraphs make up columns of text and columns of text make up pages. When handling type we are dealing with designed letterforms. Only very few experienced typographers take the time and considerable *effort to design new typefaces. Though over the years thousands of typefaces have been designed, so the choice for us*

Quark Default H&J

	Min	Opt	Max
W ✐	85	110	250
C ✐	0	0	4
H ✓	ON	HZ 14mm	
T	0		

Individual letters make words and words make up sentences (or lines) and lines make up paragraphs. Paragraphs make up columns of text and columns of text make up pages. When handling type we are dealing with designed letterforms. Only very few experienced typographers take the time and considerable *effort to design new typefaces. Though over the years thousands of typefaces have been designed, so the choice for us*

User Defined H&J

	Min	Opt	Max
W ✐	70	90	115
C ✐	-3	0	5
H ✓	ON		
T	0		

Individual letters make words and words make up sentences (or lines) and lines make up paragraphs. Paragraphs make up columns of text and columns of text make up pages. When handling type we are dealing with designed letterforms. Only very few experienced typographers take the time and considerable effort to design new *typefaces. Though over the years thousands of typefaces have been designed, so the choice for us is enormous. When* we look

User Defined H&J

	Min	Opt	Max
W ✐	75	100	115
C ✐	0	0	10
H ✓	ON		
T ✓	2		

Individual letters make words and words make up sentences (or lines) and lines make up paragraphs. Paragraphs make up columns of text and columns of text make up pages. When handling type we are dealing with designed letterforms. Only very few experienced typographers take the time and considerable *effort to design new typefaces. Though over the years thousands of typefaces have been designed, so the choice for us*

User Defined H&J

	Min	Opt	Max
W ✐	85	100	250
C ✐	0	0	0
H	OFF		
T	0		

Individual letters make words and words make up sentences (or lines) and lines make up paragraphs. Paragraphs make up columns of text and columns of text make up pages. When handling type we are dealing with designed letterforms. Only very few experienced typographers take the time and considerable effort to design new typefaces. Though over the years thousands of typefaces have been designed, so the choice for us is enormous. When we look at individual typefaces we can see that they each have their own particular characteristics and "personality." We may examine each of the 26 letters of the alphabet and can see that their form and structure has a special relationship with the background on *which it sits. Not only does the shape of the letter itself give it form, but the space around it accents its individuality. When two or more letters are put side by side they create words which are recognised as complete units, rather than collections of*

User Defined H&J

	Min	Opt	Max
W ✐	90	100	120
C ✐	0	3	10
H	OFF		
T	0		

Individual letters make words and words make up sentences (or lines) and lines make up paragraphs. Paragraphs make up columns of text and columns of text make up pages. When handling type we are dealing with designed letterforms. Only very few experienced typographers take the time and considerable effort to design new typefaces. Though over the years thousands of typefaces have been designed, so the choice for us is enormous. When we look at individual typefaces we can see that they each have their own particular characteristics and "personality." We may examine each of the 26 letters of the *alphabet and can see that their form and structure has a special relationship with the background on which it sits. Not only does the shape of the letter itself give it form, but the space around it accents its*

ITALIC *AaBbCcDdEeFfGgHhIiJjKkLlMmNnOoPpQqRrSsTt UuVvWwXxYyZz1234567890;:'"!?@#£$&*()+=*

BOLD **AaBbCcDdEeFfGgHhIiJjKkLlMmNnOoPpQqRrSsTt UuVvWwXxYyZz1234567890;:'"!?@#£$&*()+=**

BOLD ITALIC ***AaBbCcDdEeFfGgHhIiJjKkLlMmNnOoPpQqRrSsTt UuVvWwXxYyZz1234567890;:'"!?@#£$&*()+=***

EXPERT REGULAR A B CDD E F $^{1}/_{4}$G $^{1}/_{2}$H $^{3}/_{4}$I $^{1}/_{8}$J $^{3}/_{8}$K $^{5}/_{8}$L $^{7}/_{8}$M $^{1}/_{3}$N $^{2}/_{3}$O P Q ^{r}R S T uffvfiwflxffiyfflz1234567890;: ‼ ¢3$&.. ^{0} .—

EXPERT ITALIC $^{1}/_{4}$ $^{1}/_{2}$ $^{3}/_{4}$ $^{1}/_{8}$ $^{3}/_{8}$ $^{5}/_{8}$ $^{7}/_{8}$ $^{1}/_{3}$ $^{2}/_{3}$ *ff fi fl ffi ffl 1234567890;: ¢3$ .. ^{0} .—*

EXPERT BOLD $^{1}/_{4}$ $^{1}/_{2}$ $^{3}/_{4}$ $^{1}/_{8}$ $^{3}/_{8}$ $^{5}/_{8}$ $^{7}/_{8}$ $^{1}/_{3}$ $^{2}/_{3}$ **ff fi fl ffi ffl 1234567890;: ¢3$ .. ^{0} .—**

EXPERT BOLD ITALIC $^{1}/_{4}$ $^{1}/_{2}$ $^{3}/_{4}$ $^{1}/_{8}$ $^{3}/_{8}$ $^{5}/_{8}$ $^{7}/_{8}$ $^{1}/_{3}$ $^{2}/_{3}$ ***ff fi fl ffi ffl 1234567890;: ¢3$ .. ^{0} .—***

SERIF

Garamond

Monotype Garamond Bold

The italic is a spirited, elegant script-like form with a marked angle to it. The tail of the capital "Q," has a flourish, as does the leg of the lowercase "k," the strokes of the wide lowercase "x," the strong curve of the lowercase "y" and the exuberant finish of the lowercase "z." The curve of the lowercase "p" also overshoots the stem.

The family now includes regular and bold weights, with italics. The Expert face includes many different ligatures and swashes.

From left to right each group consists of letters set in Monotype, Adobe, and original Garamond.

Setting Tips

Garamond's readability does not benefit from excessive tracking. Leading can be less than the standard 120% used by most software as the relatively small x-height maintains a strong horizontal emphasis.

RRR

hhh

ggg

AaBbCc

ABCDEFGHIJKLMNOPQRSTUVWXYZ
abcdefghijklmnopqrstuvwxyz1234567890;:'"!?@#£$&*()+=

Quark Default H&J

	Min	Opt	Max
W	85	110	250
C	0	0	4
H ✓	ON		
T	0		

Individual letters make words and words make up sentences (or lines) and lines make up paragraphs. Paragraphs make up columns of text and columns of text make up pages. When handling type we are dealing with designed letterforms. Only very few experienced typographers take the time and considerable effort to design new typefaces. Though over the years thousands of typefaces have been designed, so the choice for us is enormous. When we look at individual typefaces we can see that they each have their own particular characteristics and "personality." We may *examine each of the 26 letters of the alphabet and can see that their form and structure has a special relationship with the background on which it sits. Not only does the shape of the let-*

User Defined H&J

	Min	Opt	Max
W	80	105	150
C	0	3	3
H ✓	ON		
T ✓	3		

Individual letters make words and words make up sentences (or lines) and lines make up paragraphs. Paragraphs make up columns of text and columns of text make up pages. When handling type we are dealing with designed letterforms. Only very few experienced typographers take the time and considerable effort to design new typefaces. Though over the years thousands of typefaces have been designed, so the choice for us is enormous. When we look at individual typefaces we can see that they each have their own particular characteristics and "personality." *We may examine each of the 26 letters of the alphabet and can see that their form and structure has a special relationship with the background on*

User Defined H&J

	Min	Opt	Max
W	70	100	105
C	0	0	3
H ✓	ON		
T	0		

User Defined H&J

	Min	Opt	Max
W	70	100	120
C	–3	0	5
H	OFF		
T	0		

Individual letters make words and words make up sentences (or lines) and lines make up paragraphs. Paragraphs make up columns of text and columns of text make up pages. When handling type we are dealing with designed letterforms. Only very few experienced typographers take the time and considerable effort to design new typefaces.

Individual letters make words and words make up sentences (or lines) and lines make up paragraphs. Paragraphs make up columns of text and columns of text make up pages. When handling type we are dealing with designed letterforms. Only very few experienced typographers take

Individual letters make words and words make up sentences (or lines) and lines make up paragraphs. Paragraphs make up columns of text and columns of text make up pages. When handling type we are dealing with designed letterforms. Only very few experienced typographers take *the time and considerable effort to design new typefaces. Though over the years thousands of type-*

Individual letters make words and words make up sentences (or lines) and lines make up paragraphs. Paragraphs make up columns of text and columns of text make up pages. When handling type we are dealing with designed letterforms. Only very few experienced typographers take *the time and considerable effort to design new typefaces. Though over the years thousands of*

Individual letters make words and words make up sentences (or lines) and lines make up paragraphs. Paragraphs make up columns of text and columns of text make up pages. When handling type we are dealing with designed letterforms. Only very few experienced typographers take *the time and considerable effort to design new typefaces. Though over the years thousands of*

Individual letters make words and words make up sentences (or lines) and lines make up paragraphs. Paragraphs make up columns of text and columns of text make up pages. When handling type we are dealing with designed letterforms. Only very few experienced typographers take *the time and considerable effort to design new typefaces. Though over the years thousands of*

Individual letters make words and words make up sentences (or lines) and lines make up paragraphs. Paragraphs make up columns of text and columns of text make up pages. When handling type we are dealing with designed letterforms. Only very few experienced typographers take the time and consid-*erable effort to design new typefaces. Though over the years thousands of typefaces have been*

Individual letters make words and words make up sentences (or lines) and lines make up paragraphs. Paragraphs make up columns of text and columns of text make up pages. When handling type we are dealing with designed letterforms. Only very few experienced typographers *take the time and considerable effort to design new typefaces. Though over the years thou-*

Individual letters make words and words make up sentences (or lines) and lines make up paragraphs. Paragraphs make up columns of text and columns of text make up pages. When handling type we are dealing with designed letterforms. Only very few experienced typographers take the time and considerable effort to design new typefaces. Though over the years thousands of typefaces have been designed, so the choice for us is enormous. When we look at individual typefaces we can see that they each have their own particular characteristics and "personality." We may examine each of the 26 letters of the alphabet and can see that *their form and structure has a special relationship with the background on which it sits. Not only does the shape of the letter itself give it form, but the space around it accents its individuality. When two*

Garamond

GARAMOND

–10 –8 –9

–5 5 9

Garamond Roman showing overall tracking of –5 with letter pairs individually kerned as shown. Gray text indicates 0 tracking and no kerning of letter pairs.

When we look at individual typefaces we can see that they each have their own particular characteristics and "personality."

18-pt set solid
(no extra leading),
track 0.

SERIF

Glypha

Adrian Frutiger, the renowned Swiss graphic designer and typographer, designed Glypha in 1977 for release by Linotype two years later. An ingeniously modified slab serif, Glypha has a subtlety in its stroke weight that demonstrates Frutiger's ability to give an unexpected distinction to everyday type designs. Its rhythmic feel is in total contrast to the mechanical construction of the early monoline slab serif designs. Frutiger has managed to give Glypha the same visual impact but with more style. It needs a little leading and is best suited to shorter-length text settings and display contexts, including signage.

TYPE CHARACTERISTICS

Glypha belongs to the slab serif group of typefaces, as its flat, square serifs suggest. It has comparatively condensed proportions for a face of its kind, making it suitable for use where space is limited. Glypha has the same large x-height, short ascenders and descenders as Frutiger's earlier slab serif design, Serifa, giving it high legibility and authority. The Glypha family is numerically identified and includes, 35 Thin, 45 Light, 55 Roman, 65 Bold, and 75 Black, each with its own oblique version. The thin weight is unusually thin and is really suitable only for the smoothest quality art paper.

Identifiable characteristics in the capital letters include the distinctive horizontal tail that sits on the baseline, emerging from the right of the bowl of the "Q." The lowercase "g" is single storey, the top finish of the lowercase "t" is angled and, discreetly, the lowercase "u" has half serifs to the top of the strokes.

Generally, the subtle stroke change can be seen in the curves at junctions. The clean lines and slightly condensed look allow for tighter setting without too much loss of readability.

ABCDEFGHIJKLMNOPQRSTUVWXYZ
abcdefghijklmnopqrstuvwxyz1234567890;:'"!?@#£$&*()+=

Quark Default H&J			
	Min	Opt	Max
W ✎	85	110	250
C ✎	0	0	4
H ✓	ON		
T	0		

Individual letters make words and words make up sentences (or lines) and lines make up paragraphs. Paragraphs make up columns of text and columns of text make up pages. When handling type we are dealing with designed letterforms. Only very few experienced typographers take the time and considerable effort to design new typefaces. Though over the years thousands of typefaces have been designed, so the choice for us is enormous. When we look at individual typefaces we can see that they *each have their own particular characteristics and "personality." We may examine each of the 26 letters of the alphabet and can see that their form and*

User Defined H&J			
	Min	Opt	Max
W ✎	90	100	105
C ✎	0	0	3
H ✓	ON		
T ✓	3		

Individual letters make words and words make up sentences (or lines) and lines make up paragraphs. Paragraphs make up columns of text and columns of text make up pages. When handling type we are dealing with designed letterforms. Only very few experienced typographers take the time and considerable effort to design new typefaces. Though over the years thousands of typefaces have been designed, so the choice for us is enormous. When we look at individual typefaces we can see that *they each have their own particular characteristics and "personality." We may examine each of the 26 letters of the alphabet and can see that their form*

User Defined H&J			
	Min	Opt	Max
W ✎	80	105	150
C ✎	0	3	3
H ✓	ON		
T	0		

Individual letters make words and words make up sentences (or lines) and lines make up paragraphs. Paragraphs make up columns of text and columns of text make up pages. When handling type we are dealing with designed letterforms. Only very few experienced typographers take the time and considerable effort to design new typefaces. Though over the years thousands of typefaces have been designed, so the choice for us is enormous. When we look at individual typefaces we can see that they *each have their own particular characteristics and "personality." We may examine each of the 26 letters of the alphabet and can see that their form and*

Quark Default H&J

	Min	Opt	Max
W	85	110	250
C	0	0	4
H ✓	ON		
T	0		

Individual letters make words and words make up sentences (or lines) and lines make up paragraphs. Paragraphs make up columns of text and columns of text make up pages. When handling type we are dealing with designed letterforms. Only *very few experienced typographers take the time and considerable effort to*

Quark Default H&J

	Min	Opt	Max
W	85	110	250
C	0	0	4
H	OFF		
T	0		

Individual letters make words and words make up sentences (or lines) and lines make up paragraphs. Paragraphs make up columns of text and columns of text make up pages. When handling type we are dealing with designed letterforms. Only *very few experienced typographers take the time and considerable effort to*

Quark Default H&J

	Min	Opt	Max
W	85	110	250
C	0	0	4
H ✓	ON		
T	0		

Individual letters make words and words make up sentences (or lines) and lines make up paragraphs. Paragraphs make up columns of text and columns of text make up pages. When handling type we are dealing with designed letterforms. Only *very few experienced typographers take the time and considerable effort to*

Quark Default H&J

	Min	Opt	Max
W	85	110	250
C	0	0	4
H ✓	ON HZ 14mm		
T	0		

Individual letters make words and words make up sentences (or lines) and lines make up paragraphs. Paragraphs make up columns of text and columns of text make up pages. When handling type we are dealing with designed letterforms. Only *very few experienced typographers take the time and considerable effort to*

User Defined H&J

	Min	Opt	Max
W	70	90	115
C	-3	0	5
H ✓	ON		
T	0		

Individual letters make words and words make up sentences (or lines) and lines make up paragraphs. Para-graphs make up columns of text and columns of text make up pages. When han-dling type we are dealing with designed letterforms. Only very few experienced *typographers take the time and considerable effort to design new typefaces.*

User Defined H&J

	Min	Opt	Max
W	70	110	250
C	-3	5	10
H ✓	ON		
T ✓	2		

Individual letters make words and words make up sentences (or lines) and lines make up paragraphs. Paragraphs make up columns of text and columns of text make up pages. When handling type we are dealing with designed letterforms. Only *very few experienced typographers take the time and considerable effort to*

User Defined H&J

	Min	Opt	Max
W	85	100	250
C	0	0	0
H	OFF		
T	0		

Individual letters make words and words make up sentences (or lines) and lines make up paragraphs. Paragraphs make up columns of text and columns of text make up pages. When handling type we are dealing with designed letterforms. Only very few experienced typographers take the time and considerable effort to design new typefaces. Though over the years thousands of typefaces have been designed, so the choice for us is enormous. When we look at individual typefaces we can see that they each have their own particular characteristics and *"personality." We may examine each of the 26 letters of the alphabet and can see that their form and structure has a special relationship with the background on which it*

User Defined H&J

	Min	Opt	Max
W	70	100	105
C	0	0	3
H	OFF		
T	0		

Individual letters make words and words make up sentences (or lines) and lines make up paragraphs. Paragraphs make up columns of text and columns of text make up pages. When handling type we are dealing with designed letterforms. Only very few experienced typographers take the time and considerable effort to design new typefaces. Though over the years thousands of typefaces have been designed, so the choice for us is enormous. When we look at individual typefaces we can see that they *each have their own particular characteristics and "personality." We may examine each of the 26 letters of the alphabet and can see that their form and*

ITALIC *AaBbCcDdEeFfGgHhIiJjKkLlMmNnOoPpQqRrSsTt UuVvWwXxYyZz1234567890;:'"!?@#£$&*()+=*

THIN AaBbCcDdEeFfGgHhIiJjKkLlMmNnOoPpQqRrSsTt UuVvWwXxYyZz1234567890;:'"!?@#£$&*()+=

THIN ITALIC *AaBbCcDdEeFfGgHhIiJjKkLlMmNnOoPpQqRrSsTt UuVvWwXxYyZz1234567890;:'"!?@#£$&*()+=*

LIGHT AaBbCcDdEeFfGgHhIiJjKkLlMmNnOoPpQqRrSsTt UuVvWwXxYyZz1234567890;:'"!?@#£$&*()+=

LIGHT ITALIC *AaBbCcDdEeFfGgHhIiJjKkLlMmNnOoPpQqRrSsTt UuVvWwXxYyZz1234567890;:'"!?@#£$&*()+=*

SERIF

*qrstuvwxyz*PAGES84-85

Glypha

Glypha Bold

Setting Tips

The x-height of Glypha compared with Perpetua (*see* right) shows that Glypha can have considerable relative impact at small sizes. Consider changing weight when mixing oblique with roman.

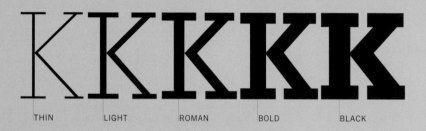

| THIN | LIGHT | ROMAN | BOLD | BLACK |

A good range of weights provides good "signposting" possibilities.

Individual letters make words and words make up sentences (or lines) and lines make up paragraphs. **Paragraphs** make up columns of text and columns of text make up pages.

AaBbC

ABCDEFGHIJKLMNOPQRSTUVWXYZ
abcdefghijklmnopqrstuvwxyz1234567890;:'"!?@#£$&*()+=

Quark Default H&J

	Min	Opt	Max
W ✎	85	110	250
C ✎	0	0	4
H ✓	ON		
T	0		

Individual letters make words and words make up sentences (or lines) and lines make up paragraphs. Paragraphs make up columns of text and columns of text make up pages. When handling type we are dealing with designed letterforms. Only very few experienced typographers take the time and considerable effort to design new typefaces. Though over the years thousands of typefaces have been designed, so the choice for us is enormous. When we look at individual typefaces we can see that *they each have their own particular characteristics and "personality." We may examine each of the 26 letters of the alphabet and can see that their form*

User Defined H&J

	Min	Opt	Max
W ✎	70	100	105
C ✎	0	0	3
H ✓	ON		
T ✓	3		

Individual letters make words and words make up sentences (or lines) and lines make up paragraphs. Paragraphs make up columns of text and columns of text make up pages. When handling type we are dealing with designed letterforms. Only very few experienced typographers take the time and considerable effort to design new typefaces. Though over the years thousands of typefaces have been designed, so the choice for us is enormous. When we look at individual typefaces *we can see that they each have their own particular characteristics and "personality." We may examine each of the 26 letters of the alphabet and*

User Defined H&J				User Defined H&J			
	Min	Opt	Max		Min	Opt	Max
W ✎	75	100	105	W ✎	70	100	105
C ✎	0	0	0	C ✎	0	0	3
H ✓	ON			H	OFF		
T	0			T	0		

Individual letters make words and words make up sentences (or lines) and lines make up paragraphs. Paragraphs make up columns of text and columns of text make up pages. When handling type we are dealing with designed letterforms. Only very few experienced typographers take the time and con-

Individual letters make words and words make up sentences (or lines) and lines make up paragraphs. Paragraphs make up columns of text and columns of text make up pages. When handling type we are dealing with designed letterforms. Only very few experienced typographers take the time and

Quark Default H&J			
	Min	Opt	Max
W	85	110	250
C	0	0	4
H ✓	ON		
T	0		

Individual letters make words and words make up sentences (or lines) and lines make up paragraphs. Paragraphs make up columns of text and columns of text make up pages. When handling type we are dealing with designed letterforms. *Only very few experienced typographers take the time and considerable*

Quark Default H&J			
	Min	Opt	Max
W	85	110	250
C	0	0	4
H	OFF		
T	0		

Individual letters make words and words make up sentences (or lines) and lines make up paragraphs. Paragraphs make up columns of text and columns of text make up pages. When handling type we are dealing with designed letterforms. *Only very few experienced typographers take the time and*

Quark Default H&J			
	Min	Opt	Max
W	85	110	250
C	0	0	4
H ✓	ON		
T	0		

Individual letters make words and words make up sentences (or lines) and lines make up paragraphs. Paragraphs make up columns of text and columns of text make up pages. When handling type we are dealing with designed letterforms. *Only very few experienced typographers take the time and considerable*

Quark Default H&J			
	Min	Opt	Max
W	85	110	250
C	0	0	4
H ✓	ON HZ 14mm		
T	0		

Individual letters make words and words make up sentences (or lines) and lines make up paragraphs. Paragraphs make up columns of text and columns of text make up pages. When handling type we are dealing with designed letterforms. *Only very few experienced typographers take the time and considerable*

User Defined H&J			
	Min	Opt	Max
W	100	100	200
C	0	5	10
H ✓	ON		
T	0		

Individual letters make words and words make up sentences (or lines) and lines make up paragraphs. Paragraphs make up columns of text and columns of text make up pages. When handling type we are dealing with designed letterforms. *Only very few experienced typographers take the time and considerable*

User Defined H&J			
	Min	Opt	Max
W	100	100	200
C	0	5	10
H ✓	ON		
T ✓	2		

Individual letters make words and words make up sentences (or lines) and lines make up paragraphs. Paragraphs make up columns of text and columns of text make up pages. When handling type we are dealing with designed letterforms. *Only very few experienced typographers take the time and consider-*

User Defined H&J			
	Min	Opt	Max
W	85	100	250
C	0	0	0
H	OFF		
T	0		

Individual letters make words and words make up sentences (or lines) and lines make up paragraphs. Paragraphs make up columns of text and columns of text make up pages. When handling type we are dealing with designed letterforms. Only very few experienced typographers take the time and considerable effort to design new typefaces. Though over the years thousands of typefaces have been designed, so the choice for us is enormous. When we look at individual typefaces we can see that they each have their own particular *characteristics and "personality." We may examine each of the 26 letters of the alphabet and can see that their form and structure has a special relationship with the*

Above: Glypha slab serif (gray) compared with a typical serif font (shown as outline).

Glypha Roman showing overall tracking of −10 with letter pairs individually kerned as shown. Gray text indicates 0 tracking and no kerning of letter pairs.

BOLD **AaBbCcDdEeFfGgHhIiJjKkLlMmNnOoPpQqRrSsTt UuVvWwXxYyZz1234567890;:'"!?@#£$&*()+=**

BOLD ITALIC ***AaBbCcDdEeFfGgHhIiJjKkLlMmNnOoPpQqRrSsTt UuVvWwXxYyZz1234567890;:'"!?@#£$&*()+=***

BLACK **AaBbCcDdEeFfGgHhIiJjKkLlMmNnOoPpQqRrSsTt UuVvWwXxYyZz1234567890;:'"!?@#£$&*()+=**

BLACK ITALIC ***AaBbCcDdEeFfGgHhIiJjKkLlMmNnOoPpQqRrSsTt UuVvWwXxYyZz1234567890;:'"!?@#£$&*()+=***

−12 −3 −11

−11 −17 −3 −12 −4

SERIF

*qrstuvwxyz*PAGES86-87

Melior

Another of Hermann Zapf's successful workhorse designs, Melior was released in 1952 by the Stempel foundry. It has a straightforward, no-nonsense character specifically designed for use in newspapers and magazines, where legibility and readability are of prime importance. Typically, Zapf's genius manages to combine these practical requirements with an individual and contemporary feel. This individuality is found in the curves, which Zapf based on a new and different form, unknowingly predating what was to become known as the "super-ellipse." Interestingly, more than a decade later, this innovative form was acclaimed as a novel device designed by a Dutch architect.

TYPE CHARACTERISTICS

Although a rather heavy roman, Melior has open characters with a large x-height and short ascenders and descenders. The contrast in the stroke weight is average and the letters finish with well-defined, square-bracketed serifs that, it has been proved, make for excellent legibility and give the type a distinct color.

Melior is identifiable by its overall super-ellipse shape, similar to a rectangle with heavily rounded corners. The capital "C" and "G" are rather square and the capital "Q" has a medium-length tail that just hooks inside the bowl.

Melior comes in regular and bold weights with italic versions and a roman extra-bold weight. The italics are essentially sloped romans, with three lowercase exceptions: the "a" (single storey), the "t" (with a thin descender), and the "k" (with a square shoulder and curled finish to the leg).

Melior is a versatile, economical face that works at both text and larger sizes and combines well with Zapf's unique design, Optima. Matrix, the Scangraphic version of Zapf's Melior, has a dollar sign with a broken vertical stroke.

ABCDEFGHIJKLMNOPQRSTUVWXYZ
abcdefghijklmnopqrstuvwxyz1234567890;:'"!?@#£$&*()+=

Quark Default H&J

	Min	Opt	Max
W	85	110	250
C	0	0	4
H ✓	ON		
T	0		

Individual letters make words and words make up sentences (or lines) and lines make up paragraphs. Paragraphs make up columns of text and columns of text make up pages. When handling type we are dealing with designed letterforms. Only very few experienced typographers take the time and considerable effort to design new typefaces. Though over the years thousands of typefaces have been designed, so the choice for us is enormous. When we look at individual typefaces we can see that they each have their *own particular characteristics and "personality." We may examine each of the 26 letters of the alphabet and can see that their form and structure has a special rela-*

User Defined H&J

	Min	Opt	Max
W	90	100	120
C	0	3	10
H ✓	ON		
T ✓	3		

Individual letters make words and words make up sentences (or lines) and lines make up paragraphs. Paragraphs make up columns of text and columns of text make up pages. When handling type we are dealing with designed letterforms. Only very few experienced typographers take the time and considerable effort to design new typefaces. Though over the years thousands of typefaces have been designed, so the choice for us is enormous. When we look at individual typefaces we can see that they *each have their own particular characteristics and "personality." We may examine each of the 26 letters of the alphabet and can see that their form and struc-*

User Defined H&J

	Min	Opt	Max
W	80	105	150
C	0	3	3
H ✓	ON		
T	0		

Individual letters make words and words make up sentences (or lines) and lines make up paragraphs. Paragraphs make up columns of text and columns of text make up pages. When handling type we are dealing with designed letterforms. Only very few experienced typographers take the time and considerable effort to design new typefaces. Though over the years thousands of typefaces have been designed, so the choice for us is enormous. When we look at individual typefaces we can see that they each have their *own particular characteristics and "personality." We may examine each of the 26 letters of the alphabet and can see that their form and structure has a special*

Quark Default H&J	Min	Opt	Max
W	85	110	250
C	0	0	4
H ✓	ON		
T	0		

Individual letters make words and words make up sentences (or lines) and lines make up paragraphs. Paragraphs make up columns of text and columns of text make up pages. When handling type we are dealing with designed letterforms. Only very few experienced *typographers take the time and considerable effort to design new typefaces.*

Quark Default H&J	Min	Opt	Max
W	85	110	250
C	0	0	4
H	OFF		
T	0		

Individual letters make words and words make up sentences (or lines) and lines make up paragraphs. Paragraphs make up columns of text and columns of text make up pages. When handling type we are dealing with designed letterforms. Only very few experienced *typographers take the time and considerable effort to design new typefaces.*

Quark Default H&J	Min	Opt	Max
W	85	110	250
C	0	0	4
H ✓	ON		
T	0		

Individual letters make words and words make up sentences (or lines) and lines make up paragraphs. Paragraphs make up columns of text and columns of text make up pages. When handling type we are dealing with designed letterforms. Only *very few experienced typographers take the time and considerable effort to design*

Quark Default H&J	Min	Opt	Max
W	85	110	250
C	0	0	4
H ✓	ON	HZ 14mm	
T	0		

Individual letters make words and words make up sentences (or lines) and lines make up paragraphs. Paragraphs make up columns of text and columns of text make up pages. When handling type we are dealing with designed letterforms. Only *very few experienced typographers take the time and considerable effort to*

User Defined H&J	Min	Opt	Max
W	70	110	250
C	-3	5	10
H ✓	ON		
T	0		

Individual letters make words and words make up sentences (or lines) and lines make up paragraphs. Paragraphs make up columns of text and columns of text make up pages. When handling type we are dealing with designed letterforms. Only very few experienced typog*raphers take the time and considerable effort to design new typefaces. Though over the*

User Defined H&J	Min	Opt	Max
W	70	90	115
C	-3	0	5
H ✓	ON		
T ✓	2		

Individual letters make words and words make up sentences (or lines) and lines make up paragraphs. Paragraphs make up columns of text and columns of text make up pages. When handling type we are dealing with designed letterforms. Only very few experienced *typographers take the time and considerable effort to design new typefaces. Though*

User Defined H&J	Min	Opt	Max
W	85	100	250
C	0	0	0
H	OFF		
T	0		

Individual letters make words and words make up sentences (or lines) and lines make up paragraphs. Paragraphs make up columns of text and columns of text make up pages. When handling type we are dealing with designed letterforms. Only very few experienced typographers take the time and considerable effort to design new typefaces. Though over the years thousands of typefaces have been designed, so the choice for us is enormous. When we look at individual typefaces we can see that they each have their own particular characteristics and "personality." We may examine each of the *26 letters of the alphabet and can see that their form and structure has a special relationship with the background on which it sits. Not only does the shape of the letter itself give it*

User Defined H&J	Min	Opt	Max
W	70	100	105
C	0	0	3
H	OFF		
T	0		

Individual letters make words and words make up sentences (or lines) and lines make up paragraphs. Paragraphs make up columns of text and columns of text make up pages. When handling type we are dealing with designed letterforms. Only very few experienced typographers take the time and considerable effort to design new typefaces. Though over the years thousands of typefaces have been designed, so the choice for us is enormous. When we look at individual typefaces we can see that they each have their own particular *characteristics and "personality." We may examine each of the 26 letters of the alphabet and can see that their form and structure* has a special relationship with the

ITALIC
AaBbCcDdEeFfGgHhIiJjKkLlMmNnOoPpQqRrSsTt UuVvWwXxYyZz1234567890;:'"!?@#£$&()+=*

BOLD
AaBbCcDdEeFfGgHhIiJjKkLlMmNnOoPpQqRrSsTt UuVvWwXxYyZz1234567890;:'"!?@#£$&*()+=

BOLD ITALIC
AaBbCcDdEeFfGgHhIiJjKkLlMmNnOoPpQqRrSsTt UuVvWwXxYyZz1234567890;:'"!?@#£$&*()+=

Minion

Designed by Robert Slimbach for Adobe in 1990, Minion is an Adobe Originals typeface design issued in 1991. Although essentially a contemporary text face, Minion has its roots in the elegantly durable and eminently readable Old Style type designs. To this late Renaissance grace, Slimbach adds a more robust up-to-date functionality that, with its digital versatility, makes Minion appropriate to a wide range of design contexts, including continuous text and display settings.

In the italics, the capitals are virtually sloped romans. However, the lowercase letters have a lively, more calligraphic quality that complements the roman.

Minion is also available as a Multiple Master font, which greatly extends its versatility (*See* Multiple Masters, page 130).

TYPE CHARACTERISTICS

The stress or direction of the curved letters is slightly inclined to the left, in keeping with the Old Style group of typefaces. Among the capital letters, the "J" has a short tail, the "R" an elegant, tapered leg, and the "Q" a tail that only just crosses into the counter. However, one of Minion's main characteristics is that it displays little overall quirkiness. In the italics, the lowercase "k" has a small curl to the leg.

The Minion family includes regular, swash display, semibold, bold, and black weights. All, with the exception of black, have italics. The Minion Expert Set is comprehensive and includes ligatures and swashes for all weights.

67

AaBbCc

ABCDEFGHIJKLMNOPQRSTUVWXYZ
abcdefghijklmnopqrstuvwxyz1234567890;:'"!?@#£$&*()+=
ABCDEFGHIJKLMNOPQRSTUVWXYZ1234567890fffiflffifffl

Quark Default H&J

	Min	Opt	Max
W ✐	85	110	250
C ✐	0	0	4
H ✓	ON		
T	0		

Individual letters make words and words make up sentences (or lines) and lines make up paragraphs. Paragraphs make up columns of text and columns of text make up pages. When handling type we are dealing with designed letterforms. Only very few experienced typographers take the time and considerable effort to design new typefaces. Though over the years thousands of typefaces have been designed, so the choice for us is enormous. When we look at individual typefaces we can see that they each have their own particular characteristics and "personality." We may examine each of the 26 letters of the *alphabet and can see that their form and structure has a special relationship with the background on which it sits. Not only does the shape of the letter itself give it form, but the space around it*

User Defined H&J

	Min	Opt	Max
W ✐	70	100	125
C ✐	0	0	2
H ✓	ON		
T ✓	3		

Individual letters make words and words make up sentences (or lines) and lines make up paragraphs. Paragraphs make up columns of text and columns of text make up pages. When handling type we are dealing with designed letterforms. Only very few experienced typographers take the time and considerable effort to design new typefaces. Though over the years thousands of typefaces have been designed, so the choice for us is enormous. When we look at individual typefaces we can see that they each have their own particular characteristics and "personality." We may examine each of the 26 letters of *the alphabet and can see that their form and structure has a special relationship with the background on which it sits. Not only does the shape of the letter itself give it form, but the space*

User Defined H&J

	Min	Opt	Max
W ✐	80	105	150
C ✐	0	3	3
H ✓	ON		
T	0		

Individual letters make words and words make up sentences (or lines) and lines make up paragraphs. Paragraphs make up columns of text and columns of text make up pages. When handling type we are dealing with designed letterforms. Only very few experienced typographers take the time and considerable effort to design new typefaces. Though over the years thousands of typefaces have been designed, so the choice for us is enormous. When we look at individual typefaces we can see that they each have their own particular characteristics and "personality." We may examine each of the 26 letters of the *alphabet and can see that their form and structure has a special relationship with the background on which it sits. Not only does the shape of the letter itself give it form, but the space around it*

Column 1

Quark Default H&J

	Min	Opt	Max
W	85	110	250
C	0	0	4
H	✓ ON		
T	0		

Individual letters make words and words make up sentences (or lines) and lines make up paragraphs. Paragraphs make up columns of text and columns of text make up pages. When handling type we are dealing with designed letterforms. Only very few experienced typographers take the time and considerable *effort to design new typefaces. Though over the years thousands of typefaces have been designed, so*

User Defined H&J

	Min	Opt	Max
W	75	100	115
C	0	0	10
H	✓ ON		
T	0		

Individual letters make words and words make up sentences (or lines) and lines make up paragraphs. Paragraphs make up columns of text and columns of text make up pages. When handling type we are dealing with designed letterforms. Only very few experienced typographers take the time and considerable *effort to design new typefaces. Though over the years thousands of typefaces have been designed, so*

User Defined H&J

	Min	Opt	Max
W	70	100	120
C	-3	0	5
H	OFF		
T	0		

Individual letters make words and words make up sentences (or lines) and lines make up paragraphs. Paragraphs make up columns of text and columns of text make up pages. When handling type we are dealing with designed letterforms. Only very few experienced typographers take the time and considerable effort to design new typefaces. Though over the years thousands of typefaces have been designed, so the choice for us is enormous. When we look at individual typefaces we can see that they each have their own particular characteristics and "personality." We may examine each of the 26 letters of the *alphabet and can see that their form and structure has a special relationship with the background on which it sits. Not only does the shape of the letter itself give it form, but the space around it*

Column 2

Quark Default H&J

	Min	Opt	Max
W	85	110	250
C	0	0	4
H	OFF		
T	0		

Individual letters make words and words make up sentences (or lines) and lines make up paragraphs. Paragraphs make up columns of text and columns of text make up pages. When handling type we are dealing with designed letterforms. Only very few experienced typographers take the time and *considerable effort to design new typefaces. Though over the years thousands of typefaces have* been

User Defined H&J

	Min	Opt	Max
W	75	90	110
C	-3	-3	3
H	✓ ON		
T	✓ 2		

Individual letters make words and words make up sentences (or lines) and lines make up paragraphs. Paragraphs make up columns of text and columns of text make up pages. When handling type we are dealing with designed letterforms. Only very few experienced typographers take the time and considerable *effort to design new typefaces. Though over the years thousands of typefaces have been designed, so*

Column 3

Quark Default H&J

	Min	Opt	Max
W	85	110	250
C	0	0	4
H	✓ ON		
T	0		

Individual letters make words and words make up sentences (or lines) and lines make up paragraphs. Paragraphs make up columns of text and columns of text make up pages. When handling type we are dealing with designed letterforms. Only very few experienced typographers take *the time and considerable effort to design new typefaces. Though over the years thousands of type-*

User Defined H&J

	Min	Opt	Max
W	85	100	250
C	0	0	0
H	OFF		
T	0		

Individual letters make words and words make up sentences (or lines) and lines make up paragraphs. Paragraphs make up columns of text and columns of text make up pages. When handling type we are dealing with designed letterforms. Only very few experienced typographers take the time and considerable effort to design new typefaces. Though over the years thousands of typefaces have been designed, so the choice for us is enormous. When we look at individual typefaces we can see that they each have their own particular characteristics and "personality." We may examine each of the 26 letters of the alphabet and can see that their form and structure *has a special relationship with the background on which it sits. Not only does the shape of the letter itself give it form, but the space around it accents its individuality. When two or more letters are put side by side*

Column 4

Quark Default H&J

	Min	Opt	Max
W	85	110	250
C	0	0	4
H	✓ ON	HZ 14mm	
T	0		

Individual letters make words and words make up sentences (or lines) and lines make up paragraphs. Paragraphs make up columns of text and columns of text make up pages. When handling type we are dealing with designed letterforms. Only very few experienced typographers take *the time and considerable effort to design new typefaces. Though over the years thousands of*

Type specimens

ITALIC *AaBbCcDdEeFfGgHhIiJjKkLlMmNnOoPpQqRrSsTt UuVvWwXxYyZz1234567890;:'"!?@#£$&*()+=*

SWASH DISPLAY *A B C D E F G H I J K L M N O P Q R S T U V W X Y Z*

SEMIBOLD **AaBbCcDdEeFfGgHhIiJjKkLlMmNnOoPpQqRrSsTt UuVvWwXxYyZz1234567890;:'"!?@#£$&*()+=**

SEMIBOLD ITALIC ***AaBbCcDdEeFfGgHhIiJjKkLlMmNnOoPpQqRrSsTt UuVvWwXxYyZz1234567890;:'"!?@#£$&*()+=***

BOLD **AaBbCcDdEeFfGgHhIiJjKkLlMmNnOoPpQqRrSsTt UuVvWwXxYyZz1234567890;:'"!?@#£$&*()+=**

BOLD ITALIC ***AaBbCcDdEeFfGgHhIiJjKkLlMmNnOoPpQqRrSsTt UuVvWwXxYyZz1234567890;:'"!?@#£$&*()+=***

BLACK **AaBbCcDdEeFfGgHhIiJjKkLlMmNnOoPpQqRrSsTt UuVvWwXxYyZz1234567890;:'"!?@#£$&*()+=**

SERIF

ITC New Baskerville

This major typeface revival is a refined and updated version of the types based on the 18th-century cut by John Baskerville, the great British printer and typefounder. Originally produced in just two styles, roman and italic, this beautiful workhorse continues to rank among the most widely used typefaces around the world.

Many versions of Baskerville exist, including several versions of ITC New Baskerville issued by Linotype and other foundries under license. The revised version was designed by Matthew Carter, the renowned type designer, and combines traditional elegance with lively contrast. Although Carter prepared the design for Mergenthaler Linotype in 1978, the marketing rights were transferred to ITC in 1982.

TYPE CHARACTERISTICS

ITC New Baskerville has a graceful round letter with tapered serifs and a subtle change from thick to delicately thin in the contrasting stroke weight, which give it color, texture, and crispness. It performs and reads very well in both extended text and display settings.

Characteristics found in the capital letters include a long serif on the bottom stroke of the "E" together with a short, "tucked-in" bar which also occurs in the "F." The bar on the "G" sits low down, and a hint of pen-like flourish can be detected in the elegant, elongated tail of the "Q." Capitals in most versions of this digitized font are closely spaced.

Among the lowercase letters, the "d" has an interesting calligraphic finish to its downstroke, which is more pronounced than in earlier versions. The bar of the "e" is very high. There is a very marked contrast between the thick outer curve of the bowl and the thin-ness of the curve at the junctions of the "b," "p," and "q." The serifs of the "s" have a vertical finish to them. The swash-like "z" and beautiful italic ampersand should also be noted.

The ITC New Baskerville family includes an extended range of weights: regular, semibold, bold, and heavy, each with its own italic version. Small caps and Old Style nonlining figures are available.

ABCDEFGHIJKLMNOPQRSTUVWXYZ
abcdefghijklmnopqrstuvwxyz1234567890;:'"!?@#£$&*()+=
1234567890

Quark Default H&J

	Min	Opt	Max
W	85	110	250
C	0	0	4
H ✓	ON		
T	0		

Individual letters make words and words make up sentences (or lines) and lines make up paragraphs. Paragraphs make up columns of text and columns of text make up pages. When handling type we are dealing with designed letterforms. Only very few experienced typographers take the time and considerable effort to design new typefaces. Though over the years thousands of typefaces have been designed, so the choice for us is enormous. When we look at individual typefaces we can see that they each have their own particular characteristics and "personality." *We may examine each of the 26 letters of the alphabet and can see that their form and structure has a special relationship with the background on which it sits. Not only*

User Defined H&J

	Min	Opt	Max
W	90	100	120
C	0	3	10
H ✓	ON		
T ✓	3		

Individual letters make words and words make up sentences (or lines) and lines make up paragraphs. Paragraphs make up columns of text and columns of text make up pages. When handling type we are dealing with designed letterforms. Only very few experienced typographers take the time and considerable effort to design new typefaces. Though over the years thousands of typefaces have been designed, so the choice for us is enormous. When we look at individual typefaces we can see that they each have their own particular characteristics and "personality." *We may examine each of the 26 letters of the alphabet and can see that their form and structure has a special relationship with the back-*

User Defined H&J

	Min	Opt	Max
W	70	90	115
C	-3	0	-5
H ✓	ON		
T	0		

Individual letters make words and words make up sentences (or lines) and lines make up paragraphs. Paragraphs make up columns of text and columns of text make up pages. When handling type we are dealing with designed letterforms. Only very few experienced typographers take the time and considerable effort to design new typefaces. Though over the years thousands of typefaces have been designed, so the choice for us is enormous. When we look at individual typefaces we can see that they each have their own particular characteristics and "personality." We may *examine each of the 26 letters of the alphabet and can see that their form and structure has a special relationship with the background on which it sits. Not only does the shape of the letter itself give it*

Quark Default H&J

	Min	Opt	Max
W ✏	85	110	250
C ✏	0	0	4
H ✓	ON		
T	0		

Individual letters make words and words make up sentences (or lines) and lines make up paragraphs. Paragraphs make up columns of text and columns of text make up pages. When handling type we are dealing with designed letterforms. Only very few experienced typographers *take the time and considerable effort to design new typefaces. Though over the years thousands*

Quark Default H&J

	Min	Opt	Max
W ✏	85	110	250
C ✏	0	0	4
H	OFF		
T	0		

Individual letters make words and words make up sentences (or lines) and lines make up paragraphs. Paragraphs make up columns of text and columns of text make up pages. When handling type we are dealing with designed letterforms. Only very few experienced typographers *take the time and considerable effort to design new typefaces. Though over the years thousands*

Quark Default H&J

	Min	Opt	Max
W ✏	85	110	250
C ✏	0	0	4
H ✓	ON		
T	0		

Individual letters make words and words make up sen-tences (or lines) and lines make up paragraphs. Paragraphs make up columns of text and columns of text make up pages. When han-dling type we are dealing with designed letterforms. Only very few experienced *typographers take the time and considerable effort to design new typefaces. Though over the years*

Quark Default H&J

	Min	Opt	Max
W ✏	85	110	250
C ✏	0	0	4
H ✓	ON	HZ 14mm	
T	0		

Individual letters make words and words make up sen-tences (or lines) and lines make up paragraphs. Paragraphs make up columns of text and columns of text make up pages. When handling type we are dealing with designed letterforms. Only very few experienced *typographers take the time and considerable effort to design new typefaces. Though over the years*

User Defined H&J

	Min	Opt	Max
W ✏	75	90	90
C ✏	–3	0	3
H ✓	ON		
T	0		

Individual letters make words and words make up sentences (or lines) and lines make up paragraphs. Paragraphs make up columns of text and columns of text make up pages. When handling type we are dealing with designed let-terforms. Only very few expe-rienced typographers take the *time and considerable effort to design new typefaces. Though over the years thousands of typefaces*

User Defined H&J

	Min	Opt	Max
W ✏	70	110	250
C ✏	–3	5	10
H ✓	ON		
T ✓	2		

Individual letters make words and words make up sen-tences (or lines) and lines make up paragraphs. Para-graphs make up columns of text and columns of text make up pages. When han-dling type we are dealing with designed letterforms. Only very few experienced typog-*raphers take the time and con-siderable effort to design new typefaces. Though over the years*

User Defined H&J

	Min	Opt	Max
W ✏	85	100	250
C ✏	0	0	0
H	OFF		
T	0		

Individual letters make words and words make up sentences (or lines) and lines make up paragraphs. Paragraphs make up columns of text and columns of text make up pages. When handling type we are dealing with designed letterforms. Only very few experienced typographers take the time and considerable effort to design new typefaces. Though over the years thousands of typefaces have been designed, so the choice for us is enormous. When we look at individual typefaces we can see that they each have their own particular characteristics and "personality." We may examine each of the 26 letters of the *alphabet and can see that their form and structure has a special relationship with the background on which it sits. Not only does the shape of the letter itself give it form, but the space around it accents its*

User Defined H&J

	Min	Opt	Max
W ✏	80	105	150
C ✏	0	3	3
H	OFF		
T	0		

Individual letters make words and words make up sentences (or lines) and lines make up paragraphs. Paragraphs make up columns of text and columns of text make up pages. When handling type we are dealing with designed letterforms. Only very few experienced typographers take the time and considerable effort to design new typefaces. Though over the years thousands of typefaces have been designed, so the choice for us is enormous. When we look at individual typefaces we can see that they each have their own particular character-*istics and "personality." We may examine each of the 26 letters of the alphabet and can see that their form and structure has a special relationship with the background on which it sits. Not*

ITALIC *AaBbCcDdEeFfGgHhIiJjKkLlMmNnOoPpQqRrSsTt UuVvWwXxYyZz1234567890;:'"!?@#£$&*()+=*

BOLD **AaBbCcDdEeFfGgHhIiJjKkLlMmNnOoPpQqRrSsTt UuVvWwXxYyZz1234567890;:'"!?@#£$&*()+=**

BOLD ITALIC ***AaBbCcDdEeFfGgHhIiJjKkLlMmNnOoPpQqRrSsTt UuVvWwXxYyZz1234567890;:'"!?@#£$&*()+=***

New Baskerville

ITC New Baskerville Bold

The differences between ITC New Baskerville and Monotype Baskerville (shown in gray) are more apparent at larger sizes. Most striking is the assertive and flared leg of the lowercase "k."

ITC New Baskerville has a smooth transition where the bowl of the "q" meets the vertical stroke and has considerably deeper descenders.

The x-height of ITC New Baskerville is marginally larger and terminals are concave.

Cross-strokes are marginally bolder in ITC New Baskerville.

AaBbCc

ABCDEFGHIJKLMNOPQRSTUVWXYZ
abcdefghijklmnopqrstuvwxyz1234567890;:'"!?@#£$&*()+=
ABCDEFGHIJKLMNOPQRSTUVWXYZ1234567890

Quark Default H&J

	Min	Opt	Max
W	85	110	250
C	0	0	4
H	✓ ON		
T	0		

Individual letters make words and words make up sentences (or lines) and lines make up paragraphs. Paragraphs make up columns of text and columns of text make up pages. When handling type we are dealing with designed letterforms. Only very few experienced typographers take the time and considerable effort to design new typefaces. Though over the years thousands of typefaces have been designed, so the choice for us is enormous. When we look at individual typefaces we can see that they each have their own particular characteristics and "personality." *We may examine each of the 26 letters of the alphabet and can see that their form and structure has a special relationship with the background on which it sits.*

User Defined H&J

	Min	Opt	Max
W	75	100	105
C	0	0	0
H	✓ ON		
T	✓ 3		

Individual letters make words and words make up sentences (or lines) and lines make up paragraphs. Paragraphs make up columns of text and columns of text make up pages. When handling type we are dealing with designed letterforms. Only very few experienced typographers take the time and considerable effort to design new typefaces. Though over the years thousands of typefaces have been designed, so the choice for us is enormous. When we look at individual typefaces we can see that they each have their own particular characteristics and "personality." *We may examine each of the 26 letters of the alphabet and can see that their form and structure has a special relationship with the background on which it sits.*

User Defined H&J

	Min	Opt	Max
W	85	100	120
C	–3	0	3
H	ON		
T	0		

User Defined H&J

	Min	Opt	Max
W	90	100	120
C	0	3	10
H	OFF		
T	0		

Individual letters make words and words make up sentences (or lines) and lines make up paragraphs. Paragraphs make up columns of text and columns of text make up pages. When handling type we are dealing with designed letterforms. Only very few experienced typographers take the time and considerable effort to design new typefaces.

Individual letters make words and words make up sentences (or lines) and lines make up paragraphs. Paragraphs make up columns of text and columns of text make up pages. When handling type we are dealing with designed letterforms. Only very few experienced typographers take the time and considerable effort to

Quark Default H&J	Min	Opt	Max
W	85	110	250
C	0	0	4
H ✓	ON		
T	0		

Individual letters make words and words make up sentences (or lines) and lines make up paragraphs. Paragraphs make up columns of text and columns of text make up pages. When handling type we are dealing with designed letterforms. Only very few experienced typographers *take the time and considerable effort to design new typefaces. Though over the years thousands*

Quark Default H&J	Min	Opt	Max
W	85	110	250
C	0	0	4
H	OFF		
T	0		

Individual letters make words and words make up sentences (or lines) and lines make up paragraphs. Paragraphs make up columns of text and columns of text make up pages. When handling type we are dealing with designed letterforms. Only very few experienced typographers *take the time and considerable effort to design new typefaces. Though over the years thousands*

Quark Default H&J	Min	Opt	Max
W	85	110	250
C	0	0	4
H ✓	ON		
T	0		

Individual letters make words and words make up sentences (or lines) and lines make up paragraphs. Paragraphs make up columns of text and columns of text make up pages. When handling type we are dealing with designed letterforms. Only very few experienced *typographers take the time and considerable effort to design new typefaces. Though over the*

Quark Default H&J	Min	Opt	Max
W	85	110	250
C	0	0	4
H ✓	ON HZ 14mm		
T	0		

Individual letters make words and words make up sentences (or lines) and lines make up paragraphs. Paragraphs make up columns of text and columns of text make up pages. When handling type we are dealing with designed letterforms. Only very few experienced *typographers take the time and considerable effort to design new typefaces. Though over the*

User Defined H&J	Min	Opt	Max
W	100	100	200
C	0	5	10
H ✓	ON		
T	0		

Individual letters make words and words make up sentences (or lines) and lines make up paragraphs. Paragraphs make up columns of text and columns of text make up pages. When handling type we are dealing with designed letterforms. Only very few experienced typographers *take the time and considerable effort to design new typefaces. Though over the years thou-*

User Defined H&J	Min	Opt	Max
W	70	90	115
C	–3	0	5
H ✓	ON		
T ✓	2		

Individual letters make words and words make up sentences (or lines) and lines make up paragraphs. Paragraphs make up columns of text and columns of text make up pages. When handling type we are dealing with designed letterforms. Only very few experienced typographers take the *time and considerable effort to design new typefaces. Though over the years thousands of type-*

User Defined H&J	Min	Opt	Max
W	85	100	250
C	0	0	0
H	OFF		
T	0		

Individual letters make words and words make up sentences (or lines) and lines make up paragraphs. Paragraphs make up columns of text and columns of text make up pages. When handling type we are dealing with designed letterforms. Only very few experienced typographers take the time and considerable effort to design new typefaces. Though over the years thousands of typefaces have been designed, so the choice for us is enormous. When we look at individual typefaces we can see that they each have their own particular characteristics and "personality." We may examine each of the 26 letters of the *alphabet and can see that their form and structure has a special relationship with the background on which it sits. Not only does the shape of the letter itself give it form, but the space around it accents its*

Baskerville

–9 –4 –7 –12 –10 –12

Baskerville showing overall tracking of –2 with letter pairs individually kerned as shown. Gray text indicates 0 tracking and no kerning of letter pairs.

BASKERVILLE

–11 –7 –3 –13 –1

BASKERVILLE
BASKERVILLE
BASKERVILLE
BASKERVILLE

Baskerville Regular caps letterspaced by increasing tracking values. From top +30, +20, +10, 0

SERIF

Monotype Perpetua

Eric Gill, the originator of Perpetua, never considered himself a professional type designer, preferring to draw letters and alphabets as an end in themselves. Two of his "alphabets"—Gill Sans and his first typeface, Perpetua, became two of his most popular typefaces. The development of Perpetua involved the notable typographic expertise of Stanley Morison, Charles Malin (the French letter cutter who made the first cut of Perpetua), Frank Pierpont, and Fritz Steltzer of Monotype. Gill, a highly skilled letter cutter himself, subsequently became deeply involved in a protracted and detailed refinement of his "alphabet." Monotype constructed Gill's letterforms in a more geometric way than he liked—the uprights becoming parallel and the serifs constructed with mechanically circular brackets.

The calligraphic Felicity Italic was later cut as a companion design to Perpetua. Both designs were named after early saints. A bold italic version was added in 1959 by Monotype.

Perpetua is an open, sensitively balanced, distinctive, classical typeface that remains widely used.

TYPE CHARACTERISTICS

Although designed in the 20th century, Perpetua has the appearance of an Old Face with a similar transition from thick to thin in the stroke weight, angled stress in the curves, and bracketed serifs. It has a small x-height, with long ascenders and descenders, and crisply cut horizontal serifs, which make it a good choice for shorter texts. Perpetua has a delicate feel, but is not an essentially economical face to use.

Among the capitals, the "U" has a distinctive foot, the "K" and "R" have half serifs to the leg, the "M" is splayed with half upper serifs, and the "J" has a medium, flat curved tail. In the lowercase letters, the "b" and "q" are lack serifs, the "d" has an unusually shaped bowl, and the "f" has a angled ending, as does the arm of the "r."

Perpetua has distinctive numerals that are available in both Old Style and lining styles.

ABCDEFGHIJKLMNOPQRSTUVWXYZ
abcdefghijklmnopqrstuvwxyz1234567890;:'"!?@#£$&*()+=
ABCDEFGHIJKLMNOPQRSTUVWXYZ1234567890fffifffffffl

Quark Default H&J

	Min	Opt	Max
W	85	110	250
C	0	0	4
H	ON		
T	0		

Individual letters make words and words make up sentences (or lines) and lines make up paragraphs. Paragraphs make up columns of text and columns of text make up pages. When handling type we are dealing with designed letterforms. Only very few experienced typographers take the time and considerable effort to design new typefaces. Though over the years thousands of typefaces have been designed, so the choice for us is enormous. When we look at individual typefaces we can see that they each have their own particular characteristics and "personality." We may examine each of the 26 letters of the alphabet and can see that their form and structure has a *special relationship with the background on which it sits. Not only does the shape of the letter itself give it form, but the space around it accents its individuality. When two or more letters are put side by side they create words which*

User Defined H&J

	Min	Opt	Max
W	70	100	125
C	0	0	2
H	ON		
T	3		

Individual letters make words and words make up sentences (or lines) and lines make up paragraphs. Paragraphs make up columns of text and columns of text make up pages. When handling type we are dealing with designed letterforms. Only very few experienced typographers take the time and considerable effort to design new typefaces. Though over the years thousands of typefaces have been designed, so the choice for us is enormous. When we look at individual typefaces we can see that they each have their own particular characteristics and "personality." We may examine each of the 26 letters of the alphabet and can see that their form and structure has *a special relationship with the background on which it sits. Not only does the shape of the letter itself give it form, but the space around it accents its individuality. When two or more letters are put side by side they create words*

User Defined H&J

	Min	Opt	Max
W	90	100	120
C	0	3	10
H	ON		
T	0		

Individual letters make words and words make up sentences (or lines) and lines make up paragraphs. Paragraphs make up columns of text and columns of text make up pages. When handling type we are dealing with designed letterforms. Only very few experienced typographers take the time and considerable effort to design new typefaces. Though over the years thousands of typefaces have been designed, so the choice for us is enormous. When we look at individual typefaces we can see that they each have their own particular characteristics and "personality." We may examine each of the 26 letters of the alphabet and can see that their form and structure has *a special relationship with the background on which it sits. Not only does the shape of the letter itself give it form, but the space around it accents its individuality When two or more letters are put side by side they create words*

Individual letters make words and words make up sentences (or lines) and lines make up paragraphs. Paragraphs make up columns of text and columns of text make up pages. When handling type we are dealing with designed letterforms. Only very few experienced typographers take the time and considerable effort to design new typefaces. *Though over the years thousands of typefaces have been designed, so the choice for us is enormous. When we look at individ-*

Individual letters make words and words make up sentences (or lines) and lines make up paragraphs. Paragraphs make up columns of text and columns of text make up pages. When handling type we are dealing with designed letterforms. Only very few experienced typographers take the time and considerable effort to design new typefaces. *Though over the years thousands of typefaces have been designed, so the choice for us is enormous. When we look*

Individual letters make words and words make up sentences (or lines) and lines make up paragraphs. Paragraphs make up columns of text and columns of text make up pages. When handling type we are dealing with designed letterforms. Only very few experienced typographers take the time and considerable effort to design new typefaces. *Though over the years thousands of typefaces have been designed, so the choice for us is enormous. When we*

Quark Default H&J

	Min	Opt	Max
W ✐	85	110	250
C ✐	0	0	4
H ✓	ON HZ 14mm		
T	0		

Individual letters make words and words make up sentences (or lines) and lines make up paragraphs. Paragraphs make up columns of text and columns of text make up pages. When handling type we are dealing with designed letterforms. Only very few experienced typographers take the time and considerable effort to design new typefaces. *Though over the years thousands of typefaces have been designed, so the choice for us is*

Individual letters make words and words make up sentences (or lines) and lines make up paragraphs. Paragraphs make up columns of text and columns of text make up pages. When handling type we are dealing with designed letterforms. Only very few experienced typographers take the time and considerable effort to design new typefaces. *Though over the years thousands of typefaces have been designed, so the choice for us is enormous. When we look at individual typefaces we*

User Defined H&J

	Min	Opt	Max
W ✐	70	90	115
C ✐	-3	0	5
H ✓	ON		
T ✓	2		

Individual letters make words and words make up sentences (or lines) and lines make up paragraphs. Paragraphs make up columns of text and columns of text make up pages. When handling type we are dealing with designed letterforms. Only very few experienced typographers take the time and considerable effort to design new typefaces. Though over *the years thousands of typefaces have been designed, so the choice for us is enormous. When we look at individual type*faces

Individual letters make words and words make up sentences (or lines) and lines make up paragraphs. Paragraphs make up columns of text and columns of text make up pages. When handling type we are dealing with designed letterforms. Only very few experienced typographers take the time and considerable effort to design new typefaces. Though over the years thousands of typefaces have been designed, so the choice for us is enormous. When we look at individual typefaces we can see that they each have their own particular characteristics and "personality." We may examine each of the 26 letters of the alphabet and can see that their form and structure has a special relationship with the background on which it sits. Not only does the shape of *the letter itself give it form, but the space around it accents its individuality. When two or more letters are put side by side they create words which are recognised as complete units, rather than collections of individual shapes. Typeset words create for us a visual*

Individual letters make words and words make up sentences (or lines) and lines make up paragraphs. Paragraphs make up columns of text and columns of text make up pages. When handling type we are dealing with designed letterforms. Only very few experienced typographers take the time and considerable effort to design new typefaces. Though over the years thousands of typefaces have been designed, so the choice for us is enormous. When we look at individual typefaces we can see that they each have their own particular characteristics and "personality." We may examine each of the 26 letters of the alphabet and can see that their form and structure has a *special relationship with the background on which it sits. Not only does the shape of the letter itself give it form, but the space around it accents its individuality. When two or more letters are put side by side they create words*

ITALIC *AaBbCcDdEeFfGgHhIiJjKkLlMmNnOoPpQqRrSsTt UuVvWwXxYyZz1234567890;:'"!?@#£$&*()+=*

BOLD **AaBbCcDdEeFfGgHhIiJjKkLlMmNnOoPpQqRrSsTt UuVvWwXxYyZz1234567890;:'"!?@#£$&*()+=**

BOLD ITALIC ***AaBbCcDdEeFfGgHhIiJjKkLlMmNnOoPpQqRrSsTt UuVvWwXxYyZz1234567890;:'"!?@#£$&*()+=***

Monotype Perpetua Bold

Perpetua

Setting Tips

As with many other faces with small x-heights, Perpetua does not need to be leaded much. The calligraphic nature of the italic (below), in particular, benefits from close line spacing and no more than normal tracking values as the example below shows. It is set 18-pt solid.

Individual letters make words and words make up sentences (or lines) and lines make up paragraphs. Paragraphs make up columns of text and columns of text make up pages. When handling type we are dealing with designed letterforms.

Highly calligraphic, the lowercase italic "g" has a uniquely shaped loop and is minus an "ear," which is unusual.

LONG ASCENDERS AND DESCENDERS

AaBbCc

ABCDEFGHIJKLMNOPQRSTUVWXYZ
abcdefghijklmnopqrstuvwxyz1234567890;:'"!?@#£$&*()+=
1234567890fffiflffiffl

Quark Default H&J			
	Min	Opt	Max
W	85	110	250
C	0	0	4
H ✓	ON		
T	0		

Individual letters make words and words make up sentences (or lines) and lines make up paragraphs. Paragraphs make up columns of text and columns of text make up pages. When handling type we are dealing with designed letterforms. Only very few experienced typographers take the time and considerable effort to design new typefaces. Though over the years thousands of typefaces have been designed, so the choice for us is enormous. When we look at individual typefaces we can see that they each have their own particular characteristics and "personality." We may *examine each of the 26 letters of the alphabet and can see that their form and structure has a special relationship with the background on which it sits. Not only does the shape of the let-*

User Defined H&J			
	Min	Opt	Max
W	70	100	120
C	–3	0	5
H ✓	ON		
T ✓	3		

Individual letters make words and words make up sentences (or lines) and lines make up paragraphs. Paragraphs make up columns of text and columns of text make up pages. When handling type we are dealing with designed letterforms. Only very few experienced typographers take the time and considerable effort to design new typefaces. Though over the years thousands of typefaces have been designed, so the choice for us is enormous. When we look at individual typefaces we can see that they each have their own particular characteristics and "personality." We may *examine each of the 26 letters of the alphabet and can see that their form and structure has a special relationship with the background on which it sits. Not only does the shape of the let-*

User Defined H&J				Quark Default H&J			
	Min	Opt	Max		Min	Opt	Max
W	70	100	105	W	70	100	120
C	0	0	3	C	–3	0	5
H ✓	ON			H	OFF		
T	0			T	0		

Individual letters make words and words make up sentences (or lines) and lines make up paragraphs. Paragraphs make up columns of text and columns of text make up pages. When handling type we are dealing with designed letterforms. Only very few experienced typographers take the time and considerable effort to design new typefaces.

Individual letters make words and words make up sentences (or lines) and lines make up paragraphs. Paragraphs make up columns of text and columns of text make up pages. When handling type we are dealing with designed letterforms. Only very few experienced typographers take the time and considerable effort to design new typefaces. Though over

	Min	Opt	Max
W	85	110	250
C	0	0	4
H ✓	ON		
T	0		

Individual letters make words and words make up sentences (or lines) and lines make up paragraphs. Paragraphs make up columns of text and columns of text make up pages. When handling type we are dealing with designed letterforms. Only very few experienced typographers *take the time and considerable effort to design new typefaces. Though over the years thousands*

	Min	Opt	Max
W	85	110	250
C	0	0	4
H	OFF		
T	0		

Individual letters make words and words make up sentences (or lines) and lines make up paragraphs. Paragraphs make up columns of text and columns of text make up pages. When handling type we are dealing with designed letterforms. Only very few experienced typographers *take the time and considerable effort to design new typefaces. Though over the years thousands*

	Min	Opt	Max
W	85	110	250
C	0	0	4
H ✓	ON		
T	0		

Individual letters make words and words make up sentences (or lines) and lines make up paragraphs. Paragraphs make up columns of text and columns of text make up pages. When handling type we are dealing with designed letterforms. Only very few experienced typographers *take the time and considerable effort to design new typefaces. Though over the years thousands*

	Min	Opt	Max
W	85	110	250
C	0	0	4
H ✓	ON	HZ 14mm	
T	0		

Individual letters make words and words make up sentences (or lines) and lines make up paragraphs. Paragraphs make up columns of text and columns of text make up pages. When handling type we are dealing with designed letterforms. Only very few experienced typographers *take the time and considerable effort to design new typefaces. Though over the years thousands*

	Min	Opt	Max
W	70	90	115
C	–3	0	5
H ✓	ON		
T	0		

Individual letters make words and words make up sentences (or lines) and lines make up paragraphs. Paragraphs make up columns of text and columns of text make up pages. When handling type we are dealing with designed letterforms. Only very few experienced typographers take the time and considerable effort to design new typefaces. Though over the years thousands of typefaces have been

	Min	Opt	Max
W	100	100	200
C	0	5	10
H ✓	ON		
T ✓	2		

Individual letters make words and words make up sentences (or lines) and lines make up paragraphs. Paragraphs make up columns of text and columns of text make up pages. When handling type we are dealing with designed letterforms. Only very few experienced typographers *take the time and considerable effort to design new typefaces. Though over the years thou-*

	Min	Opt	Max
W	85	100	250
C	0	0	0
H	OFF		
T	0		

Individual letters make words and words make up sentences (or lines) and lines make up paragraphs. Paragraphs make up columns of text and columns of text make up pages. When handling type we are dealing with designed letterforms. Only very few experienced typographers take the time and considerable effort to design new typefaces. Though over the years thousands of typefaces have been designed, so the choice for us is enormous. When we look at individual typefaces we can see that they each have their own particular characteristics and "personality." We may examine each of the 26 letters of the alphabet and can see that *their form and structure has a special relationship with the background on which it sits. Not only does the shape of the letter itself give it form, but the space around it accents its individuality. When two or more*

Perpetua
-9   -8   -2

PERPETUA
-9   -7

Perpetua Regular showing overall tracking of –5 with letter pairs individually kerned as shown. Gray text indicates 0 tracking and no kerning of letter pairs.

PERPETUA
PERPETUA
PERPETUA
PERPETUA

Perpetua Regular caps letterspaced by increasing tracking values. From top: +30, +20, +10, 0.

SERIF

Scala

Scala

While working as graphic designer for the Vrendenburg Music Center in Utrecht, the Netherlands, Martin Majoor designed Scala, a seriffed typeface primarily for use in the Center's in-house print work. Majoor had an earlier involvement with an investigation into—and the development of—digital typefaces and typographic screen display.

Scala was the earliest true text face released through FontShop International in 1991 (it was reworked and re-released in 1999). In 1993, Majoor developed a sans serif version of Scala to partner his seriffed original. The designs are independent of each other in character but share a family likeness in their overall form. They combine well in editorial contexts where a controlled, coordinated change of pace is required.

Scala is a clear, extremely legible face that works flexibly as continuous text, in short passages (for which the italic version is also useful), and in subheadings and main headings.

TYPE CHARACTERISTICS

Old Style numerals come as standard with Scala but lining numerals are available in alternative "LF" fonts. Unusually, capitals "E," "F," "T," and "Z" have double serifs to their top strokes and the capital "Q" has a seriffed tail.

The lowercase "a" has a particularly small bowl while the lowercase "b" and "q," together with the capital "P" and the numerals "6" and "9" all have open bowls.

Scala does not benefit from being squeezed in by means of horizontal scaling or tight tracking. The typeface has a certain sparkle generated by the serifs, so it is preferable not to overcrowd them.

ABCDEFGHIJKLMNOPQRSTUVWXYZ
abcdefghijklmnopqrstuvwxyz1234567890;:'"!?@#£$&*()+=
ABCDEFGHIJKLMNOPQRSTUVWXYZ

Quark Default H&J

	Min	Opt	Max
W	85	110	250
C	0	0	4
H	ON		
T	0		

Individual letters make words and words make up sentences (or lines) and lines make up paragraphs. Paragraphs make up columns of text and columns of text make up pages. When handling type we are dealing with designed letterforms. Only very few experienced typographers take the time and considerable effort to design new typefaces. Though over the years thousands of typefaces have been designed, so the choice for us is enormous. When we look at individual typefaces we can see that they each have their own particular characteristics and "personality." We may *examine each of the 26 letters of the alphabet and can see that their form and structure has a special relationship with the background on which it sits. Not only does the shape of the let-*

User Defined H&J

	Min	Opt	Max
W	70	100	120
C	–3	0	5
H	ON		
T	3		

Individual letters make words and words make up sentences (or lines) and lines make up paragraphs. Paragraphs make up columns of text and columns of text make up pages. When handling type we are dealing with designed letterforms. Only very few experienced typographers take the time and considerable effort to design new typefaces. Though over the years thousands of typefaces have been designed, so the choice for us is enormous. When we look at individual typefaces we can see that they each have their own particular characteristics and "personality." *We may examine each of the 26 letters of the alphabet and can see that their form and structure has a special relationship with the background on which it sits. Not only does the*

User Defined H&J

	Min	Opt	Max
W	70	100	120
C	–3	0	5
H	ON		
T	0		

Individual letters make words and words make up sentences (or lines) and lines make up paragraphs. Paragraphs make up columns of text and columns of text make up pages. When handling type we are dealing with designed letterforms. Only very few experienced typographers take the time and considerable effort to design new typefaces. Though over the years thousands of typefaces have been designed, so the choice for us is enormous. When we look at individual typefaces we can see that they each have their own particular characteristics and "personality." We may examine each of the 26 *letters of the alphabet and can see that their form and structure has a special relationship with the background on which it sits. Not only does the shape of the letter itself give it form, but the space*

Quark Default H&J

	Min	Opt	Max
W	85	110	250
C	0	0	4
H	ON		
T	0		

Individual letters make words and words make up sentences (or lines) and lines make up paragraphs. Paragraphs make up columns of text and columns of text make up pages. When handling type we are dealing with designed letterforms. Only very few experienced typographers take *the time and considerable effort to design new typefaces. Though over the years thousands of type-*

Quark Default H&J

	Min	Opt	Max
W	85	110	250
C	0	0	4
H	OFF		
T	0		

Individual letters make words and words make up sentences (or lines) and lines make up paragraphs. Paragraphs make up columns of text and columns of text make up pages. When handling type we are dealing with designed letterforms. Only very few experienced typographers take *the time and considerable effort to design new typefaces. Though over the years thousands of*

Quark Default H&J

	Min	Opt	Max
W	85	110	250
C	0	0	4
H	ON		
T	0		

Individual letters make words and words make up sentences (or lines) and lines make up paragraphs. Paragraphs make up columns of text and columns of text make up pages. When handling type we are dealing with designed letterforms. Only very few experienced typographers take *the time and considerable effort to design new typefaces. Though over the years thousands of type-*

Quark Default H&J

	Min	Opt	Max
W	85	110	250
C	0	0	4
H	ON HZ 14mm		
T	0		

Individual letters make words and words make up sentences (or lines) and lines make up paragraphs. Paragraphs make up columns of text and columns of text make up pages. When handling type we are dealing with designed letterforms. Only very few experienced typographers take *the time and considerable effort to design new typefaces. Though over the years thousands of*

User Defined H&J

	Min	Opt	Max
W	75	90	110
C	-3	-3	3
H	ON		
T	0		

Individual letters make words and words make up sentences (or lines) and lines make up paragraphs. Paragraphs make up columns of text and columns of text make up pages. When handling type we are dealing with designed letterforms. Only very few experienced typogra-phers take the time and consid-*erable effort to design new typefaces. Though over the years thousands of typefaces have been*

User Defined H&J

	Min	Opt	Max
W	75	90	90
C	-3	0	3
H	ON		
T	2		

Individual letters make words and words make up sentences (or lines) and lines make up paragraphs. Paragraphs make up columns of text and columns of text make up pages. When handling type we are dealing with designed letter-forms. Only very few experi-enced typographers take the *time and considerable effort to design new typefaces. Though over the years thousands of typefaces*

User Defined H&J

	Min	Opt	Max
W	85	100	250
C	0	0	0
H	OFF		
T	0		

Individual letters make words and words make up sentences (or lines) and lines make up paragraphs. Paragraphs make up columns of text and columns of text make up pages. When handling type we are dealing with designed letterforms. Only very few experienced typographers take the time and considerable effort to design new typefaces. Though over the years thousands of typefaces have been designed, so the choice for us is enormous. When we look at individual typefaces we can see that they each have their own particular characteristics and "personality." We may examine each of the 26 letters of the alphabet and can see that *their form and structure has a special relationship with the background on which it sits. Not only does the shape of the letter itself give it form, but the space around it accents its individuality. When two or more*

User Defined H&J

	Min	Opt	Max
W	70	100	120
C	-1	0	3
H	OFF		
T	0		

Individual letters make words and words make up sentences (or lines) and lines make up paragraphs. Paragraphs make up columns of text and columns of text make up pages. When handling type we are dealing with designed letterforms. Only very few experienced typographers take the time and considerable effort to design new typefaces. Though over the years thousands of typefaces have been designed, so the choice for us is enormous. When we look at individual typefaces we can see that they each have their own particular characteristics and "personality." We may *examine each of the 26 letters of the alphabet and can see that their form and structure has a special relationship with the background on which it sits. Not only does the shape of the letter*

ITALIC *AaBbCcDdEeFfGgHhIiJjKkLlMmNnOoPpQqRrSsTt UuVvWwXxYyZz1234567890;:'"!?@#£$&*()+=*

BOLD **AaBbCcDdEeFfGgHhIiJjKkLlMmNnOoPpQqRrSsTt UuVvWwXxYyZz1234567890;:'"!?@#£$&*()+=**

Times New Roman

In 1931, the Monotype Corporation produced a custom typeface for the London-based *The Times* newspaper, under the supervision of Stanley Morison. Though it was Morison's concept, the *Times* draftsman, Victor Lardent, worked Morison's sketch models through the many revisions to the final design. In his concept for Times New Roman, Morison ingeniously combined historical inspiration with modernity to produce a functional design with character that communicated clearly and directly. He said: "*The Times*, as a newspaper in a class by itself, needed not a general trade type, however good, but a face whose strength of line, firmness of contour, and economy of space fulfilled the specific editorial needs of *The Times*."

TYPE CHARACTERISTICS

Compared with earlier newspaper fonts, Times New Roman is more condensed and robust, with greater visual contrast. The Monotype version is a somewhat blacker, more faithful reproduction than some other Times Roman fonts, and so sits comfortably with contrasting typefaces. With its small, sharply cut serifs, short ascenders and descenders, and capitals matched to ascender height, the design of Times New Roman makes for visual continuity, easy reading, and excellent legibility. However, text cannot be too tightly spaced without losing this legibility. Times New Roman remains a popular choice of serif typeface for newspapers and magazines, and is one of the system fonts on both Macintosh and PC Windows, owing to its legibility on screen.

Particular characteristics of the capital letters include a high crossbar on the "G" and the midway junction of the "P." Among the lowercase letters, the "b" has a top-heavy bowl, the heaviest part of the "c" and "e" are very low down, and the "g" has a wide tail to it. The lowercase "p" and "q" follow the style of the lowercase "b" with top and bottom heavy bowls respectively, and thin junctions. The lowercase serifs are angled. A font of nonlining (Old Style) numerals and small caps is available.

ABCDEFGHIJKLMNOPQRSTUVWXYZ
abcdefghijklmnopqrstuvwxyz1234567890;:'""!?@#£$&*()+=

Quark Default H&J

		Min	Opt	Max
W	✎	85	110	250
C	✎	0		4
H	✓	ON		
T		0		

Individual letters make words and words make up sentences (or lines) and lines make up paragraphs. Paragraphs make up columns of text and columns of text make up pages. When handling type we are dealing with designed letterforms. Only very few experienced typographers take the time and considerable effort to design new typefaces. Though over the years thousands of typefaces have been designed, so the choice for us is enormous. When we look at individual typefaces we can see that they each have their own particular characteristics and "personality." We may examine each of the 26 letters of the *alphabet and can see that their form and structure has a special relationship with the background on which it sits. Not only does the shape of the letter itself give it form, but the*

User Defined H&J

		Min	Opt	Max
W	✎	90	100	120
C	✎	0	3	10
H	✓	ON		
T	✓	3		

Individual letters make words and words make up sentences (or lines) and lines make up paragraphs. Paragraphs make up columns of text and columns of text make up pages. When handling type we are dealing with designed letterforms. Only very few experienced typographers take the time and considerable effort to design new typefaces. Though over the years thousands of typefaces have been designed, so the choice for us is enormous. When we look at individual typefaces we can see that they each have their own particular characteristics and "personality." We may *examine each of the 26 letters of the alphabet and can see that their form and structure has a special relationship with the background on which it sits. Not only does the*

User Defined H&J

		Min	Opt	Max
W	✎	70	100	105
C	✎	0	0	3
H	✓	ON		
T		0		

Individual letters make words and words make up sentences (or lines) and lines make up paragraphs. Paragraphs make up columns of text and columns of text make up pages. When handling type we are dealing with designed letterforms. Only very few experienced typographers take the time and considerable effort to design new typefaces. Though over the years thousands of typefaces have been designed, so the choice for us is enormous. When we look at individual typefaces we can see that they each have their own particular characteristics and "personality." We may examine each of the 26 letters of the *alphabet and can see that their form and structure has a special relationship with the background on which it sits. Not only does the shape of the letter itself give it form, but the space*

	Min	Opt	Max
W	85	110	250
C	0	0	4
H ✓	ON		
T	0		

Individual letters make words and words make up sentences (or lines) and lines make up paragraphs. Paragraphs make up columns of text and columns of text make up pages. When handling type we are dealing with designed letterforms. Only very few experienced typographers take the time and considerable *effort to design new typefaces. Though over the years thousands of typefaces have been*

Quark Default H&J

	Min	Opt	Max
W	85	110	250
C	0	0	4
H	OFF		
T	0		

Individual letters make words and words make up sentences (or lines) and lines make up paragraphs. Paragraphs make up columns of text and columns of text make up pages. When handling type we are dealing with designed letterforms. Only very few experienced typographers take the time and *considerable effort to design new typefaces. Though over the years thousands of typefaces*

Quark Default H&J

	Min	Opt	Max
W	85	110	250
C	0	0	4
H ✓	ON		
T	0		

Individual letters make words and words make up sentences (or lines) and lines make up paragraphs. Paragraphs make up columns of text and columns of text make up pages. When handling type we are dealing with designed letterforms. Only very few experienced typographers take the time and *considerable effort to design new typefaces. Though over the years thousands of typefaces*

Quark Default H&J

	Min	Opt	Max
W	85	110	250
C	0	0	4
H ✓	ON HZ 14mm		
T	0		

Individual letters make words and words make up sentences (or lines) and lines make up paragraphs. Paragraphs make up columns of text and columns of text make up pages. When handling type we are dealing with designed letterforms. Only very few experienced typographers take the time and *considerable effort to design new typefaces. Though over the years thousands of typefaces*

User Defined H&J

	Min	Opt	Max
W	70	110	250
C	-3	5	10
H ✓	ON		
T	0		

Individual letters make words and words make up sentences (or lines) and lines make up paragraphs. Paragraphs make up columns of text and columns of text make up pages. When handling type we are dealing with designed letterforms. Only very few experienced typographers take the time and considerable *effort to design new typefaces. Though over the years thousands of typefaces have been designed,*

User Defined H&J

	Min	Opt	Max
W	75	100	115
C	0	0	10
H ✓	ON		
T ✓	2		

Individual letters make words and words make up sentences (or lines) and lines make up paragraphs. Paragraphs make up columns of text and columns of text make up pages. When handling type we are dealing with designed letterforms. Only very few experienced typographers take the time and consid-*erable effort to design new typefaces. Though over the years thousands of typefaces*

User Defined H&J

	Min	Opt	Max
W	85	100	250
C	0	0	0
H	OFF		
T	0		

Individual letters make words and words make up sentences (or lines) and lines make up paragraphs. Paragraphs make up columns of text and columns of text make up pages. When handling type we are dealing with designed letterforms. Only very few experienced typographers take the time and considerable effort to design new typefaces. Though over the years thousands of typefaces have been designed, so the choice for us is enormous. When we look at individual typefaces we can see that they each have their own particular characteristics and "personality." We may examine each of the 26 letters of the alphabet and can see that their form and structure *has a special relationship with the background on which it sits. Not only does the shape of the letter itself give it form, but the space around it accents its individuality. When two or more letters are put*

User Defined H&J

	Min	Opt	Max
W	70	100	120
C	-3	0	5
H	OFF		
T	0		

Individual letters make words and words make up sentences (or lines) and lines make up paragraphs. Paragraphs make up columns of text and columns of text make up pages. When handling type we are dealing with designed letterforms. Only very few experienced typographers take the time and considerable effort to design new typefaces. Though over the years thousands of typefaces have been designed, so the choice for us is enormous. When we look at individual typefaces we can see that they each have their own particular characteristics and "personality." We may examine each of the 26 letters of *the alphabet and can see that their form and structure has a special relationship with the background on which it sits. Not only does the shape of the letter itself give it form, but the*

ITALIC	*AaBbCcDdEeFfGgHhIiJjKkLlMmNnOoPpQqRrSsTt UuVvWwXxYyZz1234567890;:'"!?@#£$&*()+=*
BOLD	**AaBbCcDdEeFfGgHhIiJjKkLlMmNnOoPpQqRrSsTt UuVvWwXxYyZz1234567890;:'"!?@#£$&*()+=**
BOLD ITALIC	***AaBbCcDdEeFfGgHhIiJjKkLlMmNnOoPpQqRrSsTt UuVvWwXxYyZz1234567890;:'"!?@#£$&*()+=***
EXPERT REGULAR	A B CDD E F¼G½H¾I⅛J⅜K⅝L⅞M⅓N⅔O P QʳR S T uffvfiwflxffiyfflZ1234567890;: !? ¢₃$&..⁰.—
EXPERT ITALIC	¼ ½ ¾ ⅛ ⅜ ⅝ ⅞ ⅓ ⅔ *ff fi fl ffi ffl 1234567890;: ¢₃$ ..⁰.—*
EXPERT BOLD	¼ ½ ¾ ⅛ ⅜ ⅝ ⅞ ⅓ ⅔ **ff fi fl ffi ffl 1234567890;: ¢₃$ ..⁰.—**
EXPERT BOLD ITALIC	¼ ½ ¾ ⅛ ⅜ ⅝ ⅞ ⅓ ⅔ ***ff fi fl ffi ffl 1234567890;: ¢₃$ ..⁰.—***

Bliss

Bliss is a clear, sans serif typeface with an essentially English feel, designed by Jeremy Tankard in 1996. It is a versatile face that can be used equally well at text and display sizes, in signage contexts, and at a range of screen resolutions. This new design was inspired by Edward Johnston's observational notes on the "essential forms" of the alphabet, which describe the fundamentals that are key to the successful design of each letterform. Tankard says of Bliss: "...it was intended to create the first commercial typeface with an English feel since Gill Sans." The design has a marginally condensed character but an open feel to it, making it highly readable.

The forms of the capitals are based on the balanced proportions of the Roman inscriptional letter. The lowercase letters have the same natural flow and combination of straight and angled stroke ends. The rule for stroke endings is for curved forms to have a straight cut at the top and an oblique at the bottom, whereas straight forms are oblique at the top and straight at the bottom.

The italic design is derived from both sloped and cursive forms.

TYPE CHARACTERISTICS

The junctions in the roman and italic are discrete, sharp in the former and curving in the latter. The ascenders are slightly taller than the cap height, making for economical leading.

In the italic, the capitals, derived from the sloped romans, have a greater incline. The written or cursive form is the basis of the lowercase alphabet, with the exception of the lowercase "a" and "e." They are sloped versions, included to prevent the setting from appearing too soft. The more cursive forms of the lowercase "f" and "g" counterbalance this feature and lend continuity to the flow.

The family includes light, regular, medium, bold, extra bold, and heavy, all with corresponding italics. Fonts include West European accented and special characters and are available for Macintosh and PC in PostScript, Type 1, and TrueType formats.

Bliss works well in wide and narrow column widths and is particularly useful in hierarchical listing contexts.

ABCDEFGHIJKLMNOPQRSTUVWXYZ
abcdefghijklmnopqrstuvwxyz1234567890;:'"!?@#£$&*()+=
ABCDEFGHIJKLMNOPQRSTUVWXYZ1234567890

Quark Default H&J

	Min	Opt	Max
W ✓	85	110	250
C ✓	0	0	4
H ✓	ON		
T	0		

Individual letters make words and words make up sentences (or lines) and lines make up paragraphs. Paragraphs make up columns of text and columns of text make up pages. When handling type we are dealing with designed letterforms. Only very few experienced typographers take the time and considerable effort to design new typefaces. Though over the years thousands of typefaces have been designed, so the choice for us is enormous. When we look at individual typefaces we can see that they each have their own particular characteristics and "personality." We may examine each of the 26 letters of *the alphabet and can see that their form and structure has a special relationship with the background on which it sits. Not only does the shape of the letter itself give it form, but the*

User Defined H&J

	Min	Opt	Max
W ✓	70	100	125
C ✓	0	0	2
H ✓	ON		
T ✓	3		

Individual letters make words and words make up sentences (or lines) and lines make up paragraphs. Paragraphs make up columns of text and columns of text make up pages. When handling type we are dealing with designed letterforms. Only very few experienced typographers take the time and considerable effort to design new typefaces. Though over the years thousands of typefaces have been designed, so the choice for us is enormous. When we look at individual typefaces we can see that they each have their own particular characteristics and "personality." We may examine *each of the 26 letters of the alphabet and can see that their form and structure has a special relationship with the background on which it sits. Not only does the shape of the letter*

User Defined H&J

	Min	Opt	Max
W ✓	70	100	120
C ✓	0	0	6
H ✓	ON		
T	0		

Individual letters make words and words make up sentences (or lines) and lines make up paragraphs. Paragraphs make up columns of text and columns of text make up pages. When handling type we are dealing with designed letterforms. Only very few experienced typographers take the time and considerable effort to design new typefaces. Though over the years thousands of typefaces have been designed, so the choice for us is enormous. When we look at individual typefaces we can see that they each have their own particular characteristics and "personality." We may examine each of the 26 letters of *the alphabet and can see that their form and structure has a special relationship with the background on which it sits. Not only does the shape of the letter itself give it form, but the*

Quark Default H&J

	Min	Opt	Max
W	85	110	250
C	0	0	4
H ✓	ON		
T	0		

Individual letters make words and words make up sentences (or lines) and lines make up paragraphs. Paragraphs make up columns of text and columns of text make up pages. When handling type we are dealing with designed letterforms. Only very few experienced typographers take the time and considerable *effort to design new typefaces. Though over the years thousands of typefaces have been designed,*

Quark Default H&J

	Min	Opt	Max
W	85	110	250
C	0	0	4
H	OFF		
T	0		

Individual letters make words and words make up sentences (or lines) and lines make up paragraphs. Paragraphs make up columns of text and columns of text make up pages. When handling type we are dealing with designed letterforms. Only very few experienced typographers take the time and *considerable effort to design new typefaces. Though over the years thousands of typefaces*

Quark Default H&J

	Min	Opt	Max
W	85	110	250
C	0	0	4
H ✓	ON		
T	0		

Individual letters make words and words make up sentences (or lines) and lines make up paragraphs. Paragraphs make up columns of text and columns of text make up pages. When handling type we are dealing with designed letterforms. Only very few experienced typographers take *the time and considerable effort to design new typefaces. Though over the years thousands of*

Quark Default H&J

	Min	Opt	Max
W	85	110	250
C	0	0	4
H ✓	ON	HZ 14mm	
T	0		

Individual letters make words and words make up sentences (or lines) and lines make up paragraphs. Paragraphs make up columns of text and columns of text make up pages. When handling type we are dealing with designed letterforms. Only very few experienced typographers take *the time and considerable effort to design new typefaces. Though over the years thousands of*

User Defined H&J

	Min	Opt	Max
W	85	100	120
C	-3	0	3
H ✓	ON		
T	0		

Individual letters make words and words make up sentences (or lines) and lines make up paragraphs. Paragraphs make up columns of text and columns of text make up pages. When handling type we are dealing with designed letterforms. Only very few experienced typographers take the time and considerable *effort to design new typefaces. Though over the years thousands of typefaces have been designed,*

User Defined H&J

	Min	Opt	Max
W	75	90	90
C	-3	0	3
H ✓	ON		
T ✓	2		

Individual letters make words and words make up sentences (or lines) and lines make up paragraphs. Paragraphs make up columns of text and columns of text make up pages. When handling type we are dealing with designed letterforms. Only very few experienced typographers take the time and considerable *effort to design new typefaces. Though over the years thousands of typefaces have been designed,*

User Defined H&J

	Min	Opt	Max
W	85	100	250
C	0	0	0
H	OFF		
T	0		

Individual letters make words and words make up sentences (or lines) and lines make up paragraphs. Paragraphs make up columns of text and columns of text make up pages. When handling type we are dealing with designed letterforms. Only very few experienced typographers take the time and considerable effort to design new typefaces. Though over the years thousands of typefaces have been designed, so the choice for us is enormous. When we look at individual typefaces we can see that they each have their own particular characteristics and "personality." We may examine each of the 26 letters of the alphabet and can see that their form and *structure has a special relationship with the background on which it sits. Not only does the shape of the letter itself give it form, but the space around it accents its individuality. When two or more letters are put side*

User Defined H&J

	Min	Opt	Max
W	70	100	120
C	-3	0	5
H	OFF		
T	0		

Individual letters make words and words make up sentences (or lines) and lines make up paragraphs. Paragraphs make up columns of text and columns of text make up pages. When handling type we are dealing with designed letterforms. Only very few experienced typographers take the time and considerable effort to design new typefaces. Though over the years thousands of typefaces have been designed, so the choice for us is enormous. When we look at individual typefaces we can see that they each have their own particular characteristics and "personality." We may examine each of the 26 letters of *the alphabet and can see that their form and structure has a special relationship with the background on which it sits. Not only does the shape of the letter itself give it form, but the space*

ITALIC	*AaBbCcDdEeFfGgHhIiJjKkLlMmNnOoPpQqRrSsTt UuVvWwXxYyZz1234567890;:'"!?@#£$&*()+=*
LIGHT	AaBbCcDdEeFfGgHhIiJjKkLlMmNnOoPpQqRrSsTt UuVvWwXxYyZz1234567890;:'"!?@#£$&*()+=
LIGHT ITALIC	*AaBbCcDdEeFfGgHhIiJjKkLlMmNnOoPpQqRrSsTt UuVvWwXxYyZz1234567890;:'"!?@#£$&*()+=*
MEDIUM	AaBbCcDdEeFfGgHhIiJjKkLlMmNnOoPpQqRrSsTt UuVvWwXxYyZz1234567890;:'"!?@#£$&*()+=
MEDIUM ITALIC	*AaBbCcDdEeFfGgHhIiJjKkLlMmNnOoPpQqRrSsTt UuVvWwXxYyZz1234567890;:'"!?@#£$&*()+=*

Bliss

Bliss Bold

Setting Tips

As with other faces with many variants of weight, Bliss requires extra care when considering changing tracking values, as tracking for one weight may not be appropriate for another.

Bliss has cap heights significantly lower than the tops of ascenders, like many of the more recently designed sans faces.

| LIGHT | REGULAR | MEDIUM | BOLD | EXTRA BOLD | HEAVY |

The wide range of weights, as well as the range of small caps, makes Bliss suitable for heavily annotated texts with hierarchical headings and cross-referencing.

AaBbCc

ABCDEFGHIJKLMNOPQRSTUVWXYZ
abcdefghijklmnopqrstuvwxyz1234567890;:'"!?@#£$&*()+=
ABCDEFGHIJKLMNOPQRSTUVWXYZ1234567890

Quark Default H&J

		Min	Opt	Max
W		85	110	250
C		0	0	4
H	✓	ON		
T		0		

Individual letters make words and words make up sentences (or lines) and lines make up paragraphs. Paragraphs make up columns of text and columns of text make up pages. When handling type we are dealing with designed letterforms. Only very few experienced typographers take the time and considerable effort to design new typefaces. Though over the years thousands of typefaces have been designed, so the choice for us is enormous. When we look at individual typefaces we can see that they each have their own particular characteristics and "personality." We *may examine each of the 26 letters of the alphabet and can see that their form and structure has a special relationship with the background on which it sits. Not only does the shape*

User Defined H&J

		Min	Opt	Max
W		70	100	120
C		–3	0	5
H	✓	ON		
T	✓	3		

Individual letters make words and words make up sentences (or lines) and lines make up paragraphs. Paragraphs make up columns of text and columns of text make up pages. When handling type we are dealing with designed letterforms. Only very few experienced typographers take the time and considerable effort to design new typefaces. Though over the years thousands of typefaces have been designed, so the choice for us is enormous. When we look at individual typefaces we can see that they each have their own particular characteristics and "personality." *We may examine each of the 26 letters of the alphabet and can see that their form and structure has a special relationship with the background on which it sits. Not only does*

User Defined H&J

		Min	Opt	Max
W		70	100	125
C		0	0	2
H	✓	ON		
T		0		

User Defined H&J

		Min	Opt	Max
W		70	100	120
C		–1	0	3
H		OFF		
T		0		

Individual letters make words and words make up sentences (or lines) and lines make up paragraphs. Paragraphs make up columns of text and columns of text make up pages. When handling type we are dealing with designed letterforms. Only very few experienced typographers take the time and considerable effort to design new typefaces.

Individual letters make words and words make up sentences (or lines) and lines make up paragraphs. Paragraphs make up columns of text and columns of text make up pages. When handling type we are dealing with designed letterforms. Only very few experienced typographers take the time and considerable effort to design new typefaces. Though over

Quark Default H&J

	Min	Opt	Max
W	85	110	250
C	0	0	4
H	✓ ON		
T	0		

Individual letters make words and words make up sentences (or lines) and lines make up paragraphs. Paragraphs make up columns of text and columns of text make up pages. When handling type we are dealing with designed letterforms. Only very few experienced typographers take *the time and considerable effort to design new typefaces. Though over the years thou-*

Quark Default H&J

	Min	Opt	Max
W	85	110	250
C	0	0	4
H	OFF		
T	0		

Individual letters make words and words make up sentences (or lines) and lines make up paragraphs. Paragraphs make up columns of text and columns of text make up pages. When handling type we are dealing with designed letterforms. Only very few experienced typographers *take the time and considerable effort to design new typefaces. Though over the years*

Quark Default H&J

	Min	Opt	Max
W	85	110	250
C	0	0	4
H	✓ ON		
T	0		

Individual letters make words and words make up sentences (or lines) and lines make up paragraphs. Paragraphs make up columns of text and columns of text make up pages. When handling type we are dealing with designed letterforms. Only very few experienced typographers *take the time and considerable effort to design new typefaces. Though over the years thou-*

Quark Default H&J

	Min	Opt	Max
W	85	110	250
C	0	0	4
H	✓ ON HZ 14mm		
T	0		

Individual letters make words and words make up sentences (or lines) and lines make up paragraphs. Paragraphs make up columns of text and columns of text make up pages. When handling type we are dealing with designed letterforms. Only very few experienced typographers *take the time and considerable effort to design new typefaces. Though over the years*

User Defined H&J

	Min	Opt	Max
W	75	90	90
C	-3	0	3
H	✓ ON		
T	0		

Individual letters make words and words make up sentences (or lines) and lines make up paragraphs. Paragraphs make up columns of text and columns of text make up pages. When handling type we are dealing with designed letterforms. Only very few experienced typographers take the time and considerable *effort to design new typefaces. Though over the years thousands of typefaces have been designed,*

User Defined H&J

	Min	Opt	Max
W	70	90	115
C	-3	0	5
H	✓ ON		
T	✓ 2		

Individual letters make words and words make up sentences (or lines) and lines make up paragraphs. Paragraphs make up columns of text and columns of text make up pages. When handling type we are dealing with designed letterforms. Only very few experienced typographers take the *time and considerable effort to design new typefaces. Though over the years thousands of*

User Defined H&J

	Min	Opt	Max
W	85	100	250
C	0	0	0
H	OFF		
T	0		

Individual letters make words and words make up sentences (or lines) and lines make up paragraphs. Paragraphs make up columns of text and columns of text make up pages. When handling type we are dealing with designed letterforms. Only very few experienced typographers take the time and considerable effort to design new typefaces. Though over the years thousands of typefaces have been designed, so the choice for us is enormous. When we look at individual typefaces we can see that they each have their own particular characteristics and "personality." We may examine each of the 26 letters of the alphabet and can see *that their form and structure has a special relationship with the background on which it sits. Not only does the shape of the letter itself give it form, but the space around it accents its individuality.*

-10 -10 -10 -10

-12 -7 -10 -10

The upper example shows Bliss with automatic pair kerning and overall tracking of −10. Gray text indicates 0 tracking and no kerning of letter pairs. The lower example shows Bliss Small Caps with no tracking and additional pair kerning.

BOLD
AaBbCcDdEeFfGgHhIiJjKkLlMmNnOoPpQqRrSsTt UuVvWwXxYyZz1234567890;:'"!?@#£$&*()+=

BOLD ITALIC
AaBbCcDdEeFfGgHhIiJjKkLlMmNnOoPpQqRrSsTt UuVvWwXxYyZz1234567890;:'"!?@#£$&()+=*

EXTRA BOLD
AaBbCcDdEeFfGgHhIiJjKkLlMmNnOoPpQqRrSsTt UuVvWwXxYyZz1234567890;:'"!?@#£$&*()+=

EXTRA BOLD ITALIC
AaBbCcDdEeFfGgHhIiJjKkLlMmNnOoPpQqRrSsTt UuVvWwXxYyZz1234567890;:'"!?@#£$&()+=*

HEAVY
AaBbCcDdEeFfGgHhIiJjKkLlMmNnOoPpQqRrSsTt UuVvWwXxYyZz1234567890;:'"!?@#£$&*()+=

HEAVY ITALIC
AaBbCcDdEeFfGgHhIiJjKkLlMmNnOoPpQqRrSsTt UuVvWwXxYyZz1234567890;:'"!?@#£$&()+=*

ITC Eras

ITC Eras is a unique, slightly inclined roman sans serif designed by Albert Boton and Albert Hollenstein in the late 1970s as an extension of the original display sizes. ITC Eras is a design of minimal change and subtle character with few idiosyncrasies to interrupt its visual flow. It has a distinctively large x-height, short ascenders and descenders, that, together with its marginal tilt, make it an individual, legible, and easily readable sans serif typeface.

TYPE CHARACTERISTICS

ITC Eras has a subtle stroke-weight change noticeable at most of the curved stroke junctions and in the slight flaring at the ends of most letter strokes (note the slight taper in the descender of the lowercase "g" and "j"). The curves are slightly flattened at the top and bottom. The capital "G" is spurless. The capitals "R" and "P" have open bowls. The tail of the capital "Q" is outside the bowl and slightly curved upward. The capital "W" has crossed center-strokes except in the outline version—an uncommon characteristic in sans serif designs.

The lowercase "a" is a very distinctive shape with a flattened and unique open bowl. Another interesting feature is the backward slanting angle of the stroke ends on the lowercase "s." The tear-shaped counters of the bowls is reminiscent of the written form.

It is an easily recognizable typeface. The family includes light, book, medium, demibold, bold, and ultra weights and an outline version. It has only one style; there is no italic version.

Setting Tips

The large x-height means that Eras can be set in small type sizes (e.g., 7 point) and still maintain legibility. Its even texture needs enough leading to keep the horizontality of the lines intact.

ABCDEFGHIJKLMNOPQRSTUVWXYZ
abcdefghijklmnopqrstuvwxyz1234567890;:'"!?@#£$&*()+=

Quark Default H&J	Min	Opt	Max
W ✎	85	110	250
C ✎	0	0	4
H ✓	ON		
T	0		

Individual letters make words and words make up sentences (or lines) and lines make up paragraphs. Paragraphs make up columns of text and columns of text make up pages. When handling type we are dealing with designed letterforms. Only very few experienced typographers take the time and considerable effort to design new typefaces. Though over the years thousands of typefaces have been designed, so the choice for us is enormous. When we look at individual typefaces we can see that they each have their **own particular characteristics and "personality." We may examine each of the 26 letters of the alphabet and can see that their form and structure has a spe-**

User Defined H&J	Min	Opt	Max
W ✎	70	100	120
C ✎	–3	0	5
H ✓	ON		
T ✓	3		

Individual letters make words and words make up sentences (or lines) and lines make up paragraphs. Paragraphs make up columns of text and columns of text make up pages. When handling type we are dealing with designed letterforms. Only very few experienced typographers take the time and considerable effort to design new typefaces. Though over the years thousands of typefaces have been designed, so the choice for us is enormous. When we look at individual typefaces we can see that they each have their own **particular characteristics and "personality." We may examine each of the 26 letters of the alphabet and can see that their form and structure has a spe-**

User Defined H&J	Min	Opt	Max
W ✎	70	100	105
C ✎	0	0	3
H ✓	ON		
T	0		

Individual letters make words and words make up sentences (or lines) and lines make up paragraphs. Paragraphs make up columns of text and columns of text make up pages. When handling type we are dealing with designed letterforms. Only very few experienced typographers take the time and considerable effort to design new typefaces. Though over the years thousands of typefaces have been designed, so the choice for us is enormous. When we look at individual typefaces we can see that they each have their own particular **characteristics and "personality." We may examine each of the 26 letters of the alphabet and can see that their form and structure has a special**

Quark Default H&J

	Min	Opt	Max
W	85	110	250
C	0	0	4
H ✓	ON		
T	0		

Individual letters make words and words make up sentences (or lines) and lines make up paragraphs. Paragraphs make up columns of text and columns of text make up pages. When handling type we are dealing with designed letterforms. Only very few experienced **typographers take the time and considerable effort to design new typefaces.**

Quark Default H&J

	Min	Opt	Max
W	85	110	250
C	0	0	4
H	OFF		
T	0		

Individual letters make words and words make up sentences (or lines) and lines make up paragraphs. Paragraphs make up columns of text and columns of text make up pages. When handling type we are dealing with designed letterforms. Only very few experienced **typographers take the time and considerable effort to design new typefaces.**

Quark Default H&J

	Min	Opt	Max
W	85	110	250
C	0	0	4
H ✓	ON		
T	0		

Individual letters make words and words make up sentences (or lines) and lines make up paragraphs. Paragraphs make up columns of text and columns of text make up pages. When handling type we are dealing with designed letterforms. Only **very few experienced typographers take the time and considerable**

Quark Default H&J

	Min	Opt	Max
W	85	110	250
C	0	0	4
H ✓	ON	HZ 14mm	
T	0		

Individual letters make words and words make up sentences (or lines) and lines make up paragraphs. Paragraphs make up columns of text and columns of text make up pages. When handling type we are dealing with designed letterforms. Only **very few experienced typographers take the time and considerable**

User Defined H&J

	Min	Opt	Max
W	70	90	115
C	-3	0	5
H ✓	ON		
T	0		

Individual letters make words and words make up sentences (or lines) and lines make up paragraphs. Paragraphs make up columns of text and columns of text make up pages. When handling type we are dealing with designed letterforms. Only very few experienced typographers **take the time and considerable effort to design new typefaces. Though over the**

User Defined H&J

	Min	Opt	Max
W	70	110	250
C	-3	5	10
H ✓	ON		
T ✓	2		

Individual letters make words and words make up sentences (or lines) and lines make up paragraphs. Paragraphs make up columns of text and columns of text make up pages. When handling type we are dealing with designed letterforms. Only very few **experienced typographers take the time and considerable effort to design**

User Defined H&J

	Min	Opt	Max
W	85	100	250
C	0	0	0
H	OFF		
T	0		

Individual letters make words and words make up sentences (or lines) and lines make up paragraphs. Paragraphs make up columns of text and columns of text make up pages. When handling type we are dealing with designed letterforms. Only very few experienced typographers take the time and considerable effort to design new typefaces. Though over the years thousands of typefaces have been designed, so the choice for us is enormous. When we look at individual typefaces we can see that they each have their own particular characteristics and "personality." We may examine each of the **26 letters of the alphabet and can see that their form and structure has a special relationship with the background on which it sits. Not only does the shape of the letter**

User Defined H&J

	Min	Opt	Max
W	70	100	120
C	-1	0	3
H	OFF		
T	0		

Individual letters make words and words make up sentences (or lines) and lines make up paragraphs. Paragraphs make up columns of text and columns of text make up pages. When handling type we are dealing with designed letterforms. Only very few experienced typographers take the time and considerable effort to design new typefaces. Though over the years thousands of typefaces have been designed, so the choice for us is enormous. When we look at individual typefaces we can see that they each have their own particular **characteristics and "personality." We may examine each of the 26 letters of the alphabet and can see that their form and structure has a special**

LIGHT AaBbCcDdEeFfGgHhIiJjKkLlMmNnOoPpQqRrSsTt
UuVvWwXxYyZz1234567890;:'"!?@#£$&*()+=

MEDIUM AaBbCcDdEeFfGgHhIiJjKkLlMmNnOoPpQqRrSsTt
UuVvWwXxYyZz1234567890;:'"!?@#£$&*()+=

DEMI **AaBbCcDdEeFfGgHhIiJjKkLlMmNnOoPpQqRrSsTt
UuVvWwXxYyZz1234567890;:'"!?@#£$&*()+=**

BOLD **AaBbCcDdEeFfGgHhIiJjKkLlMmNnOoPpQqRrSsTt
UuVvWwXxYyZz1234567890;:'"!?@#£$&*()+=**

ULTRA **AaBbCcDdEeFfGgHhIiJjKkLlMmNnOoPpQqRrSsTt
UuVvWwXxYyZz1234567890;:'"!?@#£$&*()+=**

SANS SERIF

Franklin Gothic

ITC Franklin Gothic Condensed

Originally designed by Morris Fuller Benton in 1902 for American Type Founders, Franklin Gothic is a heavy sans serif display face. Within a few years, condensed and extra condensed versions were added. Much later, under license, Victor Caruso extended the range of Franklin Gothic weights to include book and medium for the ITC version released in 1980, extending its uses to text setting. In the 1990s, condensed, compressed, and extra compressed versions were added by David Berlow to facilitate difficult copyfitting and legibility requirements, making ITC Franklin Gothic more versatile.

ITC Franklin Gothic is close to the original in feel, particularly in the subtleties of the stroke weight contrast, but has rather more condensed proportions and a slightly larger x-height as well as short ascenders and descenders, making it economical and legible.

TYPE CHARACTERISTICS

ITC Franklin Gothic is a Grotesque design with a spurred capital "G" feature with an angled junction. There is a minimal difference in the stroke width.

In the capital letters, the stroke ends of the "C" are angled whereas the tail of the "J" finishes vertically. The capital "Q" has a distinctive curled tail similar to the "Q" in Benton's News and Record Gothic typefaces. The capital "R" has a straight leg. The lowercase "a" and "g" are two storied in both the roman and the italic versions—the lowercase "g" has a closed bowl and an ear that is above the top of the x-height line, which has an almost arbitrary look to it. The serif at the foot of the numeral "1" is an unusual feature in a sans face. The italic is a slanted roman.

The family includes the original regular, book, demi, which all have italic, condensed, and extra condensed versions; medium has condensed and italics; and the heavy weight has italics only. The discrete condensed and compressed versions have corresponding italics and the extra compressed is available in the demi weight but does not have an italic version.

The specimen settings are in Franklin Gothic Condensed.

ABCDEFGHIJKLMNOPQRSTUVWXYZ
abcdefghijklmnopqrstuvwxyz1234567890;:'"!?@#£$&*()+=
ABCDEFGHIJKLMNOPQRSTUVWXYZ
abcdefghijklmnopqrstuvwxyz1234567890;:'"!?@#£$&'()+=

Quark Default H&J

		Min	Opt	Max
W	✐	85	110	250
C	✐	0	0	4
H	✓	ON		
T		0		

Individual letters make words and words make up sentences (or lines) and lines make up paragraphs. Paragraphs make up columns of text and columns of text make up pages. When handling type we are dealing with designed letterforms. Only very few experienced typographers take the time and considerable effort to design new typefaces. Though over the years thousands of typefaces have been designed, so the choice for us is enormous. When we look at individual typefaces we can see that they each have their own particular characteristics and "personality." We may examine each of the 26 letters of the alphabet and can see that their form and structure has a special relationship with the background on which it sits. Not only does the shape of the letter itself give it form, but the

User Defined H&J

		Min	Opt	Max
W	✐	70	100	125
C	✐	0	0	2
H	✓	ON		
T	✓	3		

Individual letters make words and words make up sentences (or lines) and lines make up paragraphs. Paragraphs make up columns of text and columns of text make up pages. When handling type we are dealing with designed letterforms. Only very few experienced typographers take the time and considerable effort to design new typefaces. Though over the years thousands of typefaces have been designed, so the choice for us is enormous. When we look at individual typefaces we can see that they each have their own particular characteristics and "personality." We may examine each of the 26 letters of the alphabet and can see that their form and structure has a special relationship with the background on which it sits. Not only does the shape of the let-

User Defined H&J

		Min	Opt	Max
W	✐	70	100	120
C	✐	–3	0	5
H	✓	ON		
T		0		

Individual letters make words and words make up sentences (or lines) and lines make up paragraphs. Paragraphs make up columns of text and columns of text make up pages. When handling type we are dealing with designed letterforms. Only very few experienced typographers take the time and considerable effort to design new typefaces. Though over the years thousands of typefaces have been designed, so the choice for us is enormous. When we look at individual typefaces we can see that they each have their own particular characteristics and "personality." We may examine each of the 26 letters of the alphabet and can see that their form and structure has a special relationship with the background on which it sits. Not only does the shape of the letter itself give it form, but the space

	Min	Opt	Max
W	85	110	250
C	0	0	4
H ✓	ON		
T	0		

Individual letters make words and words make up sentences (or lines) and lines make up paragraphs. Paragraphs make up columns of text and columns of text make up pages. When handling type we are dealing with designed letterforms. Only very few experienced typographers take the time and considerable effort to design new typefaces. Though over the years thousands of typefaces

	Min	Opt	Max
W	85	110	250
C	0	0	4
H	OFF		
T	0		

Individual letters make words and words make up sentences (or lines) and lines make up paragraphs. Paragraphs make up columns of text and columns of text make up pages. When handling type we are dealing with designed letterforms. Only very few experienced typographers take the time and considerable effort to design new typefaces. Though over the years thousands of typefaces

	Min	Opt	Max
W	85	110	250
C	0	0	4
H ✓	ON		
T	0		

Individual letters make words and words make up sentences (or lines) and lines make up paragraphs. Paragraphs make up columns of text and columns of text make up pages. When handling type we are dealing with designed letterforms. Only very few experienced typographers take the time and considerable effort to design new typefaces. Though over the years thousands of typefaces

	Min	Opt	Max
W	85	110	250
C	0	0	4
H ✓	ON	HZ 14mm	
T	0		

Individual letters make words and words make up sentences (or lines) and lines make up paragraphs. Paragraphs make up columns of text and columns of text make up pages. When handling type we are dealing with designed letterforms. Only very few experienced typographers take the time and considerable effort to design new typefaces. Though over the years thousands of typefaces

	Min	Opt	Max
W	75	90	90
C	–3	0	3
H ✓	ON		
T	0		

Individual letters make words and words make up sentences (or lines) and lines make up paragraphs. Paragraphs make up columns of text and columns of text make up pages. When handling type we are dealing with designed letterforms. Only very few experienced typographers take the time and considerable effort to design new typefaces. Though over the years thousands of typefaces have been designed,

	Min	Opt	Max
W	70	90	115
C	–3	0	5
H ✓	ON		
T ✓	2		

Individual letters make words and words make up sentences (or lines) and lines make up paragraphs. Paragraphs make up columns of text and columns of text make up pages. When handling type we are dealing with designed letterforms. Only very few experienced typographers take the time and considerable effort to design new typefaces. Though over the years thousands of typefaces have been designed,

	Min	Opt	Max
W	85	100	250
C	0	0	0
H	OFF		
T	0		

Individual letters make words and words make up sentences (or lines) and lines make up paragraphs. Paragraphs make up columns of text and columns of text make up pages. When handling type we are dealing with designed letterforms. Only very few experienced typographers take the time and considerable effort to design new typefaces. Though over the years thousands of typefaces have been designed, so the choice for us is enormous. When we look at individual typefaces we can see that they each have their own particular characteristics and "personality." We may examine each of the 26 letters of the alphabet and can see that their form and structure has a special relationship with the background on which it sits. Not only does the shape of the letter itself give it form, but the space around it accents its individuality. When two or more letters are

	Min	Opt	Max
W	70	100	120
C	–1	0	3
H	OFF		
T	0		

Individual letters make words and words make up sentences (or lines) and lines make up paragraphs. Paragraphs make up columns of text and columns of text make up pages. When handling type we are dealing with designed letterforms. Only very few experienced typographers take the time and considerable effort to design new typefaces. Though over the years thousands of typefaces have been designed, so the choice for us is enormous. When we look at individual typefaces we can see that they each have their own particular characteristics and "personality." We may examine each of the 26 letters of the alphabet and can see that their form and structure has a special relationship with the background on which it sits. Not only does the shape of the letter itself give it form, but the space

Franklin Gothic

60/52 POINT

60/60 POINT

Franklin Gothic Condensed can withstand negative leading in display settings.

SANS SERIF

Frutiger

Designed by the leading Swiss typographer, Adrian Frutiger, this typeface has its origins in Concorde, one of his earlier sans serif designs. Following the enormous success of his 1957 design, Univers, he was commissioned to design a signage system for the new Paris airport, Charles de Gaulle. For this, rather than use Univers, Frutiger chose to create a new, modern sans serif design more in keeping with the contemporary architecture of the airport. The outcome, Roissy, named after the village next to the airport, was fresh, refined, simple, easily recognizable, and clear to read. Frutiger's inspirational craftsmanship and deep typographic knowledge added an unusual timelessness to this freshness. In 1976, Frutiger's Roissy was adopted by the Linotype foundry, extended to include text sizes and issued under the name Frutiger.

In text sizes, Frutiger has a more animated, open texture than the early sans serif designs and can broadly be compared to Edward Johnston's Humanist typeface design, Johnston Sans, for the London Underground.

TYPE CHARACTERISTICS

Frutiger is rounded in feel with strokes that have a nuance of width change identifiable in the junctions of the letters. Apart from the lowercase "s," all the terminals of the curved letters end vertically.

Characteristics of the capital letters are the flat apex of the "A," a "C" that is wide open, and a similar spur-less "G." In the capitals the "J" has a short tail curve, the "K" a single junction, the tail of the "Q" is angled, and the "S" has very slightly angled ends.

Frutiger includes a typical double-story lowercase "a" but a single-story "g." As in its capital counterpart, the lowercase "c" is open but narrower in its design. The finish to the ascender on the lowercase "t" is angled.

The Frutiger family is numerically identified, like Univers. It ranges from 45 Light to 95 Ultra Black, each with a corresponding italic. There is a roman-only condensed version in Light, Regular, Bold, Black, and Extra Black weights—47 to 87 (*see* pages 114–115).

ABCDEFGHIJKLMNOPQRSTUVWXYZ
abcdefghijklmnopqrstuvwxyz1234567890;:'"!?@#£$&*()+=

Quark Default H&J

		Min	Opt	Max
W	⬚	85	110	250
C	⬚	0	0	4
H	✓	ON		
T		0		

Individual letters make words and words make up sentences (or lines) and lines make up paragraphs. Paragraphs make up columns of text and columns of text make up pages. When handling type we are dealing with designed letterforms. Only very few experienced typographers take the time and considerable effort to design new typefaces. Though over the years thousands of typefaces have been designed, so the choice for us is enormous. When we look at individual typefaces we can see that they *each have their own particular characteristics and "personality." We may examine each of the 26 letters of the alphabet and can see that their form and*

User Defined H&J

		Min	Opt	Max
W	⬚	90	100	100
C	⬚	0	0	3
H	✓	ON		
T	✓	3		

Individual letters make words and words make up sentences (or lines) and lines make up paragraphs. Paragraphs make up columns of text and columns of text make up pages. When handling type we are dealing with designed letterforms. Only very few experienced typographers take the time and considerable effort to design new typefaces. Though over the years thousands of typefaces have been designed, so the choice for us is enormous. When we look at individual typefaces we can see that *they each have their own particular characteristics and "personality." We may examine each of the 26 letters of the alphabet and can see that their form*

User Defined H&J

		Min	Opt	Max
W	⬚	70	100	125
C	⬚	0	0	2
H	✓	ON		
T		0		

Individual letters make words and words make up sentences (or lines) and lines make up paragraphs. Paragraphs make up columns of text and columns of text make up pages. When handling type we are dealing with designed letterforms. Only very few experienced typographers take the time and considerable effort to design new typefaces. Though over the years thousands of typefaces have been designed, so the choice for us is enormous. When we look at individual typefaces we can see that they each have their own *particular characteristics and "personality." We may examine each of the 26 letters of the alphabet and can see that their form and structure has a special*

Quark Default H&J

	Min	Opt	Max
W	85	110	250
C	0	0	4
H	✓	ON	
T	0		

Individual letters make words and words make up sentences (or lines) and lines make up paragraphs. Paragraphs make up columns of text and columns of text make up pages. When handling type we are dealing with designed letterforms. *Only very few experienced typographers take the time and considerable* effort to

Quark Default H&J

	Min	Opt	Max
W	85	110	250
C	0	0	4
H		OFF	
T	0		

Individual letters make words and words make up sentences (or lines) and lines make up paragraphs. Paragraphs make up columns of text and columns of text make up pages. When handling type we are dealing with designed letterforms. *Only very few experienced typographers take the time and considerable* effort to

Quark Default H&J

	Min	Opt	Max
W	85	110	250
C	0	0	4
H	✓	ON	
T	0		

Individual letters make words and words make up sentences (or lines) and lines make up paragraphs. Paragraphs make up columns of text and columns of text make up pages. When handling type we are dealing with designed letterforms. Only *very few experienced typographers take the time and considerable effort* to

Quark Default H&J

	Min	Opt	Max
W	85	110	250
C	0	0	4
H	✓	ON	HZ 14mm
T	0		

Individual letters make words and words make up sentences (or lines) and lines make up paragraphs. Paragraphs make up columns of text and columns of text make up pages. When handling type we are dealing with designed letterforms. Only *very few experienced typographers take the time and considerable effort* to

User Defined H&J

	Min	Opt	Max
W	70	90	115
C	–3	0	5
H	✓	ON	
T	0		

Individual letters make words and words make up sentences (or lines) and lines make up paragraphs. Para-graphs make up columns of text and columns of text make up pages. When han-dling type we are dealing with designed letterforms. Only very few experienced *typographers take the time and considerable effort to design new typefaces.*

User Defined H&J

	Min	Opt	Max
W	75	90	110
C	–3	–3	3
H	✓	ON	
T	✓	2	

Individual letters make words and words make up sentences (or lines) and lines make up paragraphs. Para-graphs make up columns of text and columns of text make up pages. When han-dling type we are dealing with designed letterforms. Only very few experienced *typographers take the time and considerable effort to design new typefaces.*

User Defined H&J

	Min	Opt	Max
W	85	100	250
C	0	0	0
H		OFF	
T	0		

Individual letters make words and words make up sentences (or lines) and lines make up paragraphs. Paragraphs make up columns of text and columns of text make up pages. When handling type we are dealing with designed letterforms. Only very few experienced typographers take the time and considerable effort to design new typefaces. Though over the years thousands of typefaces have been designed, so the choice for us is enormous. When we look at individual typefaces we can see that they each have their own particular characteristics and "personality." We may *examine each of the 26 letters of the alphabet and can see that their form and structure has a special relationship with the background on which it sits. Not only does the shape of*

User Defined H&J

	Min	Opt	Max
W	90	100	120
C	0	3	10
H		OFF	
T	0		

Individual letters make words and words make up sentences (or lines) and lines make up paragraphs. Paragraphs make up columns of text and columns of text make up pages. When handling type we are dealing with designed letterforms. Only very few experienced typographers take the time and considerable effort to design new typefaces. Though over the years thousands of typefaces have been designed, so the choice for us is enormous. When we look at individual typefaces we can see that they *each have their own particular characteristics and "personality." We may examine each of the 26 letters of the alphabet and can see that their form*

ITALIC	*AaBbCcDdEeFfGgHhIiJjKkLlMmNnOoPpQqRrSsTt UuVvWwXxYyZz1234567890;:'"!?@#£$&*()+=*
LIGHT	AaBbCcDdEeFfGgHhIiJjKkLlMmNnOoPpQqRrSsTt UuVvWwXxYyZz1234567890;:'"!?@#£$&*()+=
LIGHT ITALIC	*AaBbCcDdEeFfGgHhIiJjKkLlMmNnOoPpQqRrSsTt UuVvWwXxYyZz1234567890;:'"!?@#£$&*()+=*
BOLD	**AaBbCcDdEeFfGgHhIiJjKkLlMmNnOoPpQqRrSsTt UuVvWwXxYyZz1234567890;:'"!?@#£$&*()+=**
BOLD ITALIC	***AaBbCcDdEeFfGgHhIiJjKkLlMmNnOoPpQqRrSsTt UuVvWwXxYyZz1234567890;:'"!?@#£$&*()+=***
BLACK	**AaBbCcDdEeFfGgHhIiJjKkLlMmNnOoPpQqRrSsTt UuVvWwXxYyZz1234567890;:'"!?@#£$&*()+=**
BLACK ITALIC	***AaBbCcDdEeFfGgHhIiJjKkLlMmNnOoPpQqRrSsTt UuVvWwXxYyZz1234567890;:'"!?@#£$&*()+=***

Frutiger Bold

Frutiger

Setting Tips

Frutiger has proved to be a popular choice of face where a mood of modernity without severity is required. The strong well-spaced letterforms of Frutiger can take tighter setting if needed.

letters

mots

Individual letters make words

and words make up sentences (or

Satze

lineas

lines) and lines make up

paragraphs. Paragraphs make up

Paragraphen

columnes

columns of text and columns of

text make up pages. When

pagine

handling type we are dealing

with designed letterforms.

letterforms

AaBbCc

ABCDEFGHIJKLMNOPQRSTUVWXYZ
abcdefghijklmnopqrstuvwxyz1234567890;:'"!?@#£$&*()+=

Quark Default H&J			
	Min	Opt	Max
W ✐	85	110	250
C ✐	0	0	4
H ✓	ON		
T	0		

Individual letters make words and words make up sentences (or lines) and lines make up paragraphs. Paragraphs make up columns of text and columns of text make up pages. When handling type we are dealing with designed letterforms. Only very few experienced typographers take the time and considerable effort to design new typefaces. Though over the years thousands of typefaces have been designed, so the choice for us is enormous. When we look at individual typefaces we can see that they *each have their own particular characteristics and "personality." We may examine each of the 26 letters of the alphabet and can see that their form and*

User Defined H&J			
	Min	Opt	Max
W ✐	75	100	105
C ✐	0	0	0
H ✓	ON		
T ✓	3		

Individual letters make words and words make up sentences (or lines) and lines make up paragraphs. Paragraphs make up columns of text and columns of text make up pages. When handling type we are dealing with designed letterforms. Only very few experienced typographers take the time and considerable effort to design new typefaces. Though over the years thousands of typefaces have been designed, so the choice for us is enormous. When we look at individual typefaces we can see that *they each have their own particular characteristics and "personality." We may examine each of the 26 letters of the alphabet and can see that their form*

User Defined H&J				User Defined H&J			
	Min	Opt	Max		Min	Opt	Max
W ✐	70	100	105	W ✐	90	100	120
C ✐	0	0	3	C ✐	0	3	10
H ✓	ON			H	OFF		
T	0			T	0		

Individual letters make words and words make up sentences (or lines) and lines make up paragraphs. Paragraphs make up columns of text and columns of text make up pages. When handling type we are dealing with designed letterforms. Only very few experienced typographers take the time and consid-

Individual letters make words and words make up sentences (or lines) and lines make up paragraphs. Paragraphs make up columns of text and columns of text make up pages. When handling type we are dealing with designed letterforms. Only very few experienced typographers take the time and

Individual letters make words and words make up sentences (or lines) and lines make up paragraphs. Paragraphs make up columns of text and columns of text make up pages. When handling type we are dealing with designed letterforms. Only *very few experienced typographers take the time and considerable effort to*

Individual letters make words and words make up sentences (or lines) and lines make up paragraphs. Paragraphs make up columns of text and columns of text make up pages. When handling type we are dealing with designed letterforms. Only *very few experienced typographers take the time and considerable effort to*

Individual letters make words and words make up sentences (or lines) and lines make up paragraphs. Paragraphs make up columns of text and columns of text make up pages. When handling type we are dealing with designed letterforms. Only *very few experienced typographers take the time and considerable effort to*

Individual letters make words and words make up sentences (or lines) and lines make up paragraphs. Paragraphs make up columns of text and columns of text make up pages. When handling type we are dealing with designed letterforms. Only *very few experienced typographers take the time and considerable effort to*

Individual letters make words and words make up sentences (or lines) and lines make up paragraphs. Paragraphs make up columns of text and columns of text make up pages. When handling type we are dealing with designed letterforms. Only very few experienced *typographers take the time and considerable effort to design new typefaces.*

Individual letters make words and words make up sentences (or lines) and lines make up paragraphs. Paragraphs make up columns of text and columns of text make up pages. When handling type we are dealing with designed letterforms. Only very few experienced *typographers take the time and considerable effort to design new typefaces.*

Individual letters make words and words make up sentences (or lines) and lines make up paragraphs. Paragraphs make up columns of text and columns of text make up pages. When handling type we are dealing with designed letterforms. Only very few experienced typographers take the time and considerable effort to design new typefaces. Though over the years thousands of typefaces have been designed, so the choice for us is enormous. When we look at individual typefaces we can see that they each have their own particular characteristics *and "personality." We may examine each of the 26 letters of the alphabet and can see that their form and structure has a special relationship with the background on which it*

Frutiger Regular showing overall tracking of −11 with letter pairs individually kerned as shown. Gray text indicates 0 tracking and no kerning of letter pairs.

ULTRA BLACK **AaBbCcDdEeFfGgHhIiJjKkLlMmNnOoPpQqRrSsTt UuVvWwXxYyZz1234567890;:'"!?@#£$&*()+=**

LIGHT COND AaBbCcDdEeFfGgHhIiJjKkLlMmNnOoPpQqRrSsTt UuVvWwXxYyZz1234567890;:'"!?@#£$&*()+=

COND AaBbCcDdEeFfGgHhIiJjKkLlMmNnOoPpQqRrSsTt UuVvWwXxYyZz1234567890;:'"!?@#£$&*()+=

BOLD COND **AaBbCcDdEeFfGgHhIiJjKkLlMmNnOoPpQqRrSsTt UuVvWwXxYyZz1234567890;:'"!?@#£$&*()+=**

BLACK COND **AaBbCcDdEeFfGgHhIiJjKkLlMmNnOoPpQqRrSsTt UuVvWwXxYyZz1234567890;:'"!?@#£$&*()+=**

EXTRA BLACK COND **AaBbCcDdEeFfGgHhIiJjKkLlMmNnOoPpQqRrSsTt UuVvWwXxYyZz1234567890;:'"!?@#£$&*()+=**

Futura

Designed by the German type designer Paul Renner between 1924 and 1928, Futura has become one of the most popular geometric sans serif typefaces of the 20th century. Renner was a champion of the Bauhaus ideals and philosophy that "form follows function," so logically he designed the shape of each letterform to encapsulate no more and no less than a functional, geometric purity. Optical adjustments were made to convey the visual feel of a monoline design and a greater visual balance between the upper- and lowercase alphabets, producing a typeface design with style and authority. Futura was originally released by the Bauer Foundry in a range of six weights together with an Inline, and three weights of a condensed style that work particularly well in combinations of text and display settings. Futura works best in shorter text settings and is often seen in children's books, with the preferred single-story lowercase "a."

TYPE CHARACTERISTICS

Futura has a small x-height and long ascenders and descenders, which give solid text setting a light tonal value. The capital letters are slightly smaller than the ascender height, which adds to the feel of more space between lines of text.

The monoline feel of Futura is achieved by a slight tapering of the curves at the junctions of letters, to counteract the optical illusion of the thickening of monoline connecting points.

Futura's recognizable open and geometric construction is clearly seen in the capital "C," "G," "O," and "Q." The capital "M" has splayed strokes, the centerstroke of the capital "W" meets the cap height. The lowercase "a" is single story, the terminals of the lowercase "c" finish vertically, the bottom terminal of the lowercase "e" finishes obliquely, as do both terminals of the lowercase "s."

The Futura font has a comprehensive family of weights and italics. While the transition from light to book to bold is quite subtle, a greater contrast can be achieved by combining the light and heavy weights.

ABCDEFGHIJKLMNOPQRSTUVWXYZ
abcdefghijklmnopqrstuvwxyz1234567890;:'"!?@#£$&*()+=

Quark Default H&J			
	Min	Opt	Max
W	85	110	250
C	0	0	4
H ✓	ON		
T	0		

Individual letters make words and words make up sentences (or lines) and lines make up paragraphs. Paragraphs make up columns of text and columns of text make up pages. When handling type we are dealing with designed letterforms. Only very few experienced typographers take the time and considerable effort to design new typefaces. Though over the years thousands of typefaces have been designed, so the choice for us is enormous. When we look at individual typefaces we can see that they each have their own particular characteristics and "personality." We may examine each of the 26 letters of the alphabet and can see that their form and structure has a special relationship with the background on which it sits.

User Defined H&J			
	Min	Opt	Max
W	70	100	120
C	–3	0	5
H ✓	ON		
T ✓	3		

Individual letters make words and words make up sentences (or lines) and lines make up paragraphs. Paragraphs make up columns of text and columns of text make up pages. When handling type we are dealing with designed letterforms. Only very few experienced typographers take the time and considerable effort to design new typefaces. Though over the years thousands of typefaces have been designed, so the choice for us is enormous. When we look at individual typefaces we can see that they each have their own particular characteristics and "personality." We may examine each of the 26 letters of the alphabet and can see that their form and structure has a special relationship with the background on which it sits. Not

User Defined H&J			
	Min	Opt	Max
W	70	100	120
C	–1	0	3
H ✓	ON		
T	0		

Individual letters make words and words make up sentences (or lines) and lines make up paragraphs. Paragraphs make up columns of text and columns of text make up pages. When handling type we are dealing with designed letterforms. Only very few experienced typographers take the time and considerable effort to design new typefaces. Though over the years thousands of typefaces have been designed, so the choice for us is enormous. When we look at individual typefaces we can see that they each have their own particular characteristics and "personality." We may examine each of the 26 letters of the alphabet and can see that their form and structure has a special relationship with the background on which it sits. Not only does the shape of the let-

Quark Default H&J	Min	Opt	Max
W	85	110	250
C	0	0	4
H ✓	ON		
T	0		

Individual letters make words and words make up sentences (or lines) and lines make up paragraphs. Paragraphs make up columns of text and columns of text make up pages. When handling type we are dealing with designed letterforms. Only very few experienced typographers take the time and *considerable effort to design new typefaces. Though over the years thousands of type-*

Quark Default H&J	Min	Opt	Max
W	85	110	250
C	0	0	4
H	OFF		
T	0		

Individual letters make words and words make up sentences (or lines) and lines make up paragraphs. Paragraphs make up columns of text and columns of text make up pages. When handling type we are dealing with designed letterforms. Only very few experienced typographers take the time and *considerable effort to design new typefaces. Though over the years thousands of*

Quark Default H&J	Min	Opt	Max
W	85	110	250
C	0	0	4
H ✓	ON		
T	0		

Individual letters make words and words make up sentences (or lines) and lines make up paragraphs. Paragraphs make up columns of text and columns of text make up pages. When handling type we are dealing with designed letterforms. Only very few experienced typographers *take the time and considerable effort to design new typefaces. Though over the*

Quark Default H&J	Min	Opt	Max
W	85	110	250
C	0	0	4
H ✓	ON HZ 14mm		
T	0		

Individual letters make words and words make up sentences (or lines) and lines make up paragraphs. Paragraphs make up columns of text and columns of text make up pages. When handling type we are dealing with designed letterforms. Only very few experienced typographers *take the time and considerable effort to design new typefaces. Though over the*

User Defined H&J	Min	Opt	Max
W	70	110	250
C	-3	5	10
H ✓	ON		
T	0		

Individual letters make words and words make up sentences (or lines) and lines make up paragraphs. Paragraphs make up columns of text and columns of text make up pages. When handling type we are dealing with designed letterforms. Only very few experienced typographers take the time and con-*siderable effort to design new typefaces. Though over the years thousands of typefaces* have

User Defined H&J	Min	Opt	Max
W	60	90	110
C	-5	-5	5
H ✓	ON		
T ✓	2		

Individual letters make words and words make up sentences (or lines) and lines make up paragraphs. Paragraphs make up columns of text and columns of text make up pages. When handling type we are dealing with designed letterforms. Only very few experienced typographers take the time and consid-*erable effort to design new typefaces. Though over the years thousands of typefaces have*

User Defined H&J	Min	Opt	Max
W	85	100	250
C	0	0	0
H	OFF		
T	0		

Individual letters make words and words make up sentences (or lines) and lines make up paragraphs. Paragraphs make up columns of text and columns of text make up pages. When handling type we are dealing with designed letterforms. Only very few experienced typographers take the time and considerable effort to design new typefaces. Though over the years thousands of typefaces have been designed, so the choice for us is enormous. When we look at individual typefaces we can see that they each have their own particular characteristics and "personality." We may examine each of the 26 letters of the alphabet and can see *that their form and structure has a special relationship with the background on which it sits. Not only does the shape of the letter itself give it form, but the space around it accents its individuality.*

User Defined H&J	Min	Opt	Max
W	70	100	120
C	-3	0	5
H	OFF		
T	0		

Individual letters make words and words make up sentences (or lines) and lines make up paragraphs. Paragraphs make up columns of text and columns of text make up pages. When handling type we are dealing with designed letterforms. Only very few experienced typographers take the time and considerable effort to design new typefaces. Though over the years thousands of typefaces have been designed, so the choice for us is enormous. When we look at individual typefaces we can see that they each have their own particular characteristics and "personality." We may examine each *of the 26 letters of the alphabet and can see that their form and structure has a special relationship with the background on which it sits. Not only does the shape of the*

ITALIC	*AaBbCcDdEeFfGgHhIiJjKkLlMmNnOoPpQqRrSsTt UuVvWwXxYyZz1234567890;:'"!?@#£$&*()+=*
LIGHT	AaBbCcDdEeFfGgHhIiJjKkLlMmNnOoPpQqRrSsTt UuVvWwXxYyZz1234567890;:'"!?@#£$&*()+=
LIGHT ITALIC	*AaBbCcDdEeFfGgHhIiJjKkLlMmNnOoPpQqRrSsTt UuVvWwXxYyZz1234567890;:'"!?@#£$&*()+=*
BOOK	AaBbCcDdEeFfGgHhIiJjKkLlMmNnOoPpQqRrSsTt UuVvWwXxYyZz1234567890;:'"!?@#£$&*()+=
BOOK ITALIC	*AaBbCcDdEeFfGgHhIiJjKkLlMmNnOoPpQqRrSsTt UuVvWwXxYyZz1234567890;:'"!?@#£$&*()+=*
HEAVY	**AaBbCcDdEeFfGgHhIiJjKkLlMmNnOoPpQqRrSsTt UuVvWwXxYyZz1234567890;:'"!?@#£$&*()+=**
HEAVY ITALIC	***AaBbCcDdEeFfGgHhIiJjKkLlMmNnOoPpQqRrSsTt UuVvWwXxYyZz1234567890;:'"!?@#£$&*()+=***

Futura

Futura Bold

Setting Tips

The roundness of Futura works better when normal or plus-tracking values are used. However, minus-tracking can be applied to good effect as shown in the user defined samples.

Long ascenders break up the otherwise even rhythm, adding interest to the word formations.

Though vastly different in weight it is easy to see the common structural dynamics in this example of Futura Light and Extra Bold.

GOOD

Almost circular round letters demand fairly open tracking.

AaBbCc

ABCDEFGHIJKLMNOPQRSTUVWXYZ
abcdefghijklmnopqrstuvwxyz
1234567890;:'"!?@#£$&*()+=

Quark Default H&J			
	Min	Opt	Max
W	85	110	250
C	0	0	4
H ✓	ON		
T	0		

Individual letters make words and words make up sentences (or lines) and lines make up paragraphs. Paragraphs make up columns of text and columns of text make up pages. When handling type we are dealing with designed letterforms. Only very few experienced typographers take the time and considerable effort to design new typefaces. Though over the years thousands of typefaces have been designed, so the choice for us is enormous. *When we look at individual typefaces we can see that they each have their own particular characteristics and "personality." We may*

User Defined H&J			
	Min	Opt	Max
W	70	100	110
C	–3	0	5
H ✓	ON		
T ✓	3		

Individual letters make words and words make up sentences (or lines) and lines make up paragraphs. Paragraphs make up columns of text and columns of text make up pages. When handling type we are dealing with designed letterforms. Only very few experienced typographers take the time and considerable effort to design new typefaces. Though over the years thousands of typefaces have been designed, so the choice for us is enormous. When we look *at individual typefaces we can see that they each have their own particular characteristics and "personality." We may examine each of*

User Defined H&J			
	Min	Opt	Max
W	70	90	110
C	–6	–3	0
H ✓	ON		
T	0		

User Defined H&J			
	Min	Opt	Max
W	70	85	100
C	–5	–5	0
H	OFF		
T	0		

Individual letters make words and words make up sentences (or lines) and lines make up paragraphs. Paragraphs make up columns of text and columns of text make up pages. When handling type we are dealing with designed letterforms. Only very few experienced typographers take the

Individual letters make words and words make up sentences (or lines) and lines make up paragraphs. Paragraphs make up columns of text and columns of text make up pages. When handling type we are dealing with designed letterforms. Only very few experienced typographers take the time and

Column 1

Quark Default H&J

	Min	Opt	Max
W	85	110	250
C	0	0	4
H	ON		
T	0		

Individual letters make words and words make up sentences (or lines) and lines make up paragraphs. Paragraphs make up columns of text and columns of text make up pages. When handling type we are dealing with designed *letterforms. Only very few experienced typographers take the time*

User Defined H&J

	Min	Opt	Max
W	60	90	90
C	-5	0	10
H	ON		
T	0		

Individual letters make words and words make up sentences (or lines) and lines make up paragraphs. Paragraphs make up columns of text and columns of text make up pages. When handling type we are dealing with designed letterforms. Only *very few experienced typographers take the time and considerable*

Column 2

Quark Default H&J

	Min	Opt	Max
W	85	110	250
C	0	0	4
H	OFF		
T	0		

Individual letters make words and words make up sentences (or lines) and lines make up paragraphs. Paragraphs make up columns of text and columns of text make up pages. When handling type we are dealing with designed *letterforms. Only very few experienced typographers take the*

User Defined H&J

	Min	Opt	Max
W	75	100	115
C	0	0	10
H	ON		
T	2		

Individual letters make words and words make up sentences (or lines) and lines make up paragraphs. Paragraphs make up columns of text and columns of text make up pages. When handling type we are dealing with designed letterforms. Only very *few experienced typographers take the time*

Column 3

Quark Default H&J

	Min	Opt	Max
W	85	110	250
C	0	0	4
H	ON		
T	0		

Individual letters make words and words make up sentences (or lines) and lines make up paragraphs. Paragraphs make up columns of text and columns of text make up pages. When handling type we are *dealing with designed letterforms. Only very few experienced typog-*

User Defined H&J

	Min	Opt	Max
W	85	100	250
C	0	0	0
H	OFF		
T	0		

Individual letters make words and words make up sentences (or lines) and lines make up paragraphs. Paragraphs make up columns of text and columns of text make up pages. When handling type we are dealing with designed letterforms. Only very few experienced typographers take the time and considerable effort to design new typefaces. Though over the years thousands of typefaces have been designed, so the choice for us is enormous. When we look at individual typefaces we can see that they *each have their own particular characteristics and "personality." We may examine each of the 26 letters of the alphabet and can see that their form*

Column 4

Quark Default H&J

	Min	Opt	Max
W	85	110	250
C	0	0	4
H	ON	HZ 14mm	
T	0		

Individual letters make words and words make up sentences (or lines) and lines make up paragraphs. Paragraphs make up columns of text and columns of text make up pages. When handling type we are *dealing with designed letterforms. Only very few experienced*

BOLD ITALIC	*AaBbCcDdEeFfGgHhIiJjKkLlMmNn OoPpQqRrSsTtUuVvWwXxYyZz 1234567890;:'"!?@#£$&*()+=*
EXTRA BOLD	**AaBbCcDdEeFfGgHhIiJjKkLlMmNn OoPpQqRrSsTtUuVvWwXxYyZz 1234567890;:'"!?@#£$&*()+=**
EXTRA BOLD ITALIC	***AaBbCcDdEeFfGgHhIiJjKkLlMmNn OoPpQqRrSsTtUuVvWwXxYyZz 1234567890;:'"!?@#£$&*()+=***
COND	AaBbCcDdEeFfGgHhIiJjKkLlMmNnOoPpQqRrSsTt UuVvWwXxYyZz1234567890;:'"!?@#£$&*()+=
COND ITALIC	*AaBbCcDdEeFfGgHhIiJjKkLlMmNnOoPpQqRrSsTt UuVvWwXxYyZz1234567890;:'"!?@#£$&*()+=*

COND LIGHT	AaBbCcDdEeFfGgHhIiJjKkLlMmNnOoPpQqRrSsTt UuVvWwXxYyZz1234567890;:'"!?@#£$&*()+=
COND LIGHT ITALIC	*AaBbCcDdEeFfGgHhIiJjKkLlMmNnOoPpQqRrSsTt UuVvWwXxYyZz1234567890;:'"!?@#£$&*()+=*
COND BOLD	**AaBbCcDdEeFfGgHhIiJjKkLlMmNnOoPpQqRrSsTt UuVvWwXxYyZz1234567890;:'"!?@#£$&*()+=**
COND BOLD ITALIC	***AaBbCcDdEeFfGgHhIiJjKkLlMmNnOoPpQqRrSsTt UuVvWwXxYyZz1234567890;:'"!?@#£$&*()+=***
COND EXTRA BOLD	**AaBbCcDdEeFfGgHhIiJjKkLlMmNnOoPpQqRrSsTt UuVvWwXxYyZz1234567890;:'"!?@#£$&*()+=**
COND EXTRA BOLD ITALIC	***AaBbCcDdEeFfGgHhIiJjKkLlMmNnOoPpQqRrSsTt UuVvWwXxYyZz1234567890;:'"!?@#£$&*()+=***

Gill Sans

Gill Sans continues to be a favorite, widely used typeface. First issued by Monotype in 1928, Gill Sans was named after its designer, Eric Gill, primarily a stone carver and calligrapher. Previously, Edward Johnston, a well-known calligrapher and lettering artist, had designed an alphabet, Johnston Sans, as a house type for the London Underground. Gill, who was Johnston's friend, pupil, and collaborator in the London Transport alphabet, modeled certain features of his design on the Underground type.

Gill Sans was initially considered a titling face, but when released as a text face in 1930, it became freely available, and with its simple, modern look, quickly became a popular choice.

Gill Sans is a distinctive Humanist sans serif, with roots in the proportions of roman letterforms. It has an open feel with a comparatively small x-height and long descenders. It should not be used below 9 point for comfortable reading of text settings. It is most legible in short blocks of copy but can be used for complex information where space is at a premium.

TYPE CHARACTERISTICS

The Gill counter shapes vary, with lowercase "a" and "e" the smallest, and lowercase "o" the largest. This variation helps to lend color to text settings.

In the capital letters, the lower bowl of the "B" is wider than the upper bowl, the "J" has a short tail, and the "M" a high, pointed middle junction. Gill's own favorite was the capital "R," with its curved, slightly extended downstroke. The capital "W" is wide and has a flat middle junction and pointed lower junctions.

In the lowercase letters, the "a" and "g" are two-story and, together with the "t," have an atypical look for a sans serif design. The italic lowercase "k" has a single junction and, like the "l," has an angled finish to the top of the main stem.

The italics are a very different design from the roman. They are narrower and have a single story lowercase "a." In the capitals, the curves of the "C," "G," and "S" are unusually pronounced.

ABCDEFGHIJKLMNOPQRSTUVWXYZ
abcdefghijklmnopqrstuvwxyz1234567890;:'"!?@#£$&*()+=

Quark Default H&J			
	Min	Opt	Max
W ✐	85	110	250
C ✐	0	0	4
H ✓	ON		
T	0		

Individual letters make words and words make up sentences (or lines) and lines make up paragraphs. Paragraphs make up columns of text and columns of text make up pages. When handling type we are dealing with designed letterforms. Only very few experienced typographers take the time and considerable effort to design new typefaces. Though over the years thousands of typefaces have been designed, so the choice for us is enormous. When we look at individual typefaces we can see that they each have their own particular characteristics and "personality." We may examine each of the *26 letters of the alphabet and can see that their form and structure has a special relationship with the background on which it sits. Not only does the shape of the letter itself give it form, but the*

User Defined H&J			
	Min	Opt	Max
W ✐	70	90	90
C ✐	3	3	4
H ✓	ON		
T ✓	3		

Individual letters make words and words make up sentences (or lines) and lines make up paragraphs. Paragraphs make up columns of text and columns of text make up pages. When handling type we are dealing with designed letterforms. Only very few experienced typographers take the time and considerable effort to design new typefaces. Though over the years thousands of typefaces have been designed, so the choice for us is enormous. When we look at individual typefaces we can see that they each have their own particular characteristics and "personality." We may *examine each of the 26 letters of the alphabet and can see that their form and structure has a special relationship with the background on which it sits. Not only does the shape of the*

User Defined H&J			
	Min	Opt	Max
W ✐	90	100	120
C ✐	0	3	10
H ✓	ON		
T	0		

Individual letters make words and words make up sentences (or lines) and lines make up paragraphs. Paragraphs make up columns of text and columns of text make up pages. When handling type we are dealing with designed letterforms. Only very few experienced typographers take the time and considerable effort to design new typefaces. Though over the years thousands of typefaces have been designed, so the choice for us is enormous. When we look at individual typefaces we can see that they each have their own particular characteristics and "personality." We may examine each of *the 26 letters of the alphabet and can see that their form and structure has a special relationship with the background on which it sits. Not only does the shape of the letter itself give it form, but*

Quark Default H&J	Min	Opt	Max
W	85	110	250
C	0	0	4
H ✓	ON		
T	0		

Individual letters make words and words make up sentences (or lines) and lines make up paragraphs. Paragraphs make up columns of text and columns of text make up pages. When handling type we are dealing with designed letterforms. Only very few experienced typographers take the time and considerable *effort to design new typefaces. Though over the years thousands of typefaces have been designed,*

Quark Default H&J	Min	Opt	Max
W	85	110	250
C	0	0	4
H	OFF		
T	0		

Individual letters make words and words make up sentences (or lines) and lines make up paragraphs. Paragraphs make up columns of text and columns of text make up pages. When handling type we are dealing with designed letterforms. Only very few experienced typographers take the time and *considerable effort to design new typefaces. Though over the years thousands of typefaces have been*

Quark Default H&J	Min	Opt	Max
W	85	110	250
C	0	0	4
H ✓	ON		
T	0		

Individual letters make words and words make up sentences (or lines) and lines make up paragraphs. Paragraphs make up columns of text and columns of text make up pages. When handling type we are dealing with designed letterforms. Only very few experienced typographers take the time and *considerable effort to design new typefaces. Though over the years thousands of typefaces have been designed,*

Quark Default H&J	Min	Opt	Max
W	85	110	250
C	0	0	4
H ✓	ON	HZ 14mm	
T	0		

Individual letters make words and words make up sentences (or lines) and lines make up paragraphs. Paragraphs make up columns of text and columns of text make up pages. When handling type we are dealing with designed letterforms. Only very few experienced typographers take the time and *considerable effort to design new typefaces. Though over the years thousands of typefaces have been*

User Defined H&J	Min	Opt	Max
W	75	100	115
C	0	0	10
H ✓	ON		
T	0		

Individual letters make words and words make up sentences (or lines) and lines make up paragraphs. Paragraphs make up columns of text and columns of text make up pages. When handling type we are dealing with designed letterforms. Only very few experienced typographers take the time and considerable *effort to design new typefaces. Though over the years thousands of typefaces have been designed,*

User Defined H&J	Min	Opt	Max
W	60	90	90
C	–5	0	10
H ✓	ON		
T ✓	2		

Individual letters make words and words make up sentences (or lines) and lines make up paragraphs. Paragraphs make up columns of text and columns of text make up pages. When handling type we are dealing with designed letterforms. Only very few experienced typographers take the time and considerable *effort to design new typefaces. Though over the years thousands of typefaces have been designed, so*

User Defined H&J	Min	Opt	Max
W	85	100	250
C	0	0	0
H	OFF		
T	0		

Individual letters make words and words make up sentences (or lines) and lines make up paragraphs. Paragraphs make up columns of text and columns of text make up pages. When handling type we are dealing with designed letterforms. Only very few experienced typographers take the time and considerable effort to design new typefaces. Though over the years thousands of typefaces have been designed, so the choice for us is enormous. When we look at individual typefaces we can see that they each have their own particular characteristics and "personality." We may examine each of the 26 letters of the alphabet and can see that their form and *structure has a special relationship with the background on which it sits. Not only does the shape of the letter itself give it form, but the space around it accents its individuality. When two or more letters are put side by*

User Defined H&J	Min	Opt	Max
W	80	90	150
C	0	3	5
H	OFF		
T	0		

Individual letters make words and words make up sentences (or lines) and lines make up paragraphs. Paragraphs make up columns of text and columns of text make up pages. When handling type we are dealing with designed letterforms. Only very few experienced typographers take the time and considerable effort to design new typefaces. Though over the years thousands of typefaces have been designed, so the choice for us is enormous. When we look at individual typefaces we can see that they each have their own particular characteristics and "personality." We may examine each of the *26 letters of the alphabet and can see that their form and structure has a special relationship with the background on which it sits. Not only does the shape of the letter itself give it form, but*

ITALIC
AaBbCcDdEeFfGgHhIiJjKkLlMmNnOoPpQqRrSsTt
UuVvWwXxYyZz1234567890;:'"!?@#£$&()+=*

LIGHT
AaBbCcDdEeFfGgHhIiJjKkLlMmNnOoPpQqRrSsTt
UuVvWwXxYyZz1234567890;:'"!?@#£$&*()+=

LIGHT ITALIC
AaBbCcDdEeFfGgHhIiJjKkLlMmNnOoPpQqRrSsTt
UuVvWwXxYyZz1234567890;:'"!?@#£$&()+=*

BOLD
AaBbCcDdEeFfGgHhIiJjKkLlMmNnOoPpQqRrSsTt
UuVvWwXxYyZz1234567890;:'"!?@#£$&*()+=

BOLD ITALIC
AaBbCcDdEeFfGgHhIiJjKkLlMmNnOoPpQqRrSsTt
UuVvWwXxYyZz1234567890;:'"!?@#£$&*()+=

EXTRA BOLD
AaBbCcDdEeFfGgHhIiJjKkLlMmNnOoPpQqRrSsTt
UuVvWwXxYyZz1234567890;:'"!?@#£$&*()+=

ULTRA BOLD
AaBbCcDdEeFfGgHhIiJjKkLlMmNnOoPpQq
RrSsTtUuVvWwXxYyZz1234567890
;:'"!?@#£$&*()+=

SANS SERIF

Gill Sans

Gill Sans Bold

Setting Tips

Gill Sans benefits from increased tracking when
set at sizes below 12 point. Text set in bold can
look lumpy at larger sizes (12 or 14 point), but
a tighter tracking helps the legibility.

TRACK –5

TRACK 0

TRACK –11

TRACK –9

TRACK –7

legibility

legibility

legibility

Gill Sans italics contain
differently shaped
letterforms, unlike the
sloped italics of most sans
faces.

fap

fap

AaBbCc

ABCDEFGHIJKLMNOPQRSTUVWXYZ
abcdefghijklmnopqrstuvwxyz1234567890;:'"!?@#£$&*()+=

Quark Default H&J			
	Min	Opt	Max
W	85	110	250
C	0	0	4
H ✓	ON		
T	0		

**Individual letters make words and words make up
sentences (or lines) and lines make up paragraphs.
Paragraphs make up columns of text and columns of
text make up pages. When handling type we are dealing with designed letterforms. Only very few
experienced typographers take the time and considerable effort to design new typefaces. Though over
the years thousands of typefaces have been designed,
so the choice for us is enormous. When we look at
individual typefaces we can see that they each have
their own particular characteristics and "personality."
We may examine each of the 26 letters of the alphabet
*and can see that their form and structure has a special***

User Defined H&J			
	Min	Opt	Max
W	80	105	150
C	0	3	3
H ✓	ON		
T ✓	3		

**Individual letters make words and words make up
sentences (or lines) and lines make up paragraphs.
Paragraphs make up columns of text and columns
of text make up pages. When handling type we are
dealing with designed letterforms. Only very few
experienced typographers take the time and considerable effort to design new typefaces. Though
over the years thousands of typefaces have been
designed, so the choice for us is enormous. When
we look at individual typefaces we can see that
they each have their own particular characteristics
and "personality." We may examine each of the 26
*letters of the alphabet and can see that their form and***

User Defined H&J					User Defined H&J			
	Min	Opt	Max			Min	Opt	Max
W	75	90	110		W	70	100	120
C	–3	–3	3		C	–3	0	5
H ✓	ON				H	OFF		
T	0				T	0		

**Individual letters make words and words make up sen-
tences (or lines) and lines make up paragraphs. Paragraphs make up columns of text and columns of text
make up pages. When handling type we are dealing with
designed letterforms. Only very few experienced typographers take the time and considerable effort to design**

**Individual letters make words and words make up
sentences (or lines) and lines make up paragraphs.
Paragraphs make up columns of text and columns of
text make up pages. When handling type we are dealing
with designed letterforms. Only very few experienced
typographers take the time and considerable effort to**

Column 1 — Quark Default H&J

	Min	Opt	Max
W	85	110	250
C	0	0	4
H	✓ ON		
T	0		

Individual letters make words and words make up sentences (or lines) and lines make up paragraphs. Paragraphs make up columns of text and columns of text make up pages. When handling type we are dealing with designed letterforms. Only *very few experienced typographers take the time and considerable effort to design*

Column 2 — Quark Default H&J

	Min	Opt	Max
W	85	110	250
C	0	0	4
H	OFF		
T	0		

Individual letters make words and words make up sentences (or lines) and lines make up paragraphs. Paragraphs make up columns of text and columns of text make up pages. When handling type we are dealing with designed letterforms. Only *very few experienced typographers take the time and considerable effort to*

Column 3 — Quark Default H&J

	Min	Opt	Max
W	85	110	250
C	0	0	4
H	✓ ON		
T	0		

Individual letters make words and words make up sentences (or lines) and lines make up paragraphs. Paragraphs make up columns of text and columns of text make up pages. When handling type we are dealing with designed letterforms. Only *very few experienced typographers take the time and considerable effort to design*

Column 4 — Quark Default H&J

	Min	Opt	Max
W	85	110	250
C	0	0	4
H	✓ ON HZ 14mm		
T	0		

Individual letters make words and words make up sentences (or lines) and lines make up paragraphs. Paragraphs make up columns of text and columns of text make up pages. When handling type we are dealing with designed letterforms. Only *very few experienced typographers take the time and considerable effort to*

Column 1 — User Defined H&J

	Min	Opt	Max
W	70	90	115
C	-3	0	5
H	✓ ON		
T	0		

Individual letters make words and words make up sentences (or lines) and lines make up paragraphs. Paragraphs make up columns of text and columns of text make up pages. When handling type we are dealing with designed letterforms. Only *very few experienced typographers take the time and considerable effort to design new typefaces. Though*

Column 2 — User Defined H&J

	Min	Opt	Max
W	60	90	90
C	-5	0	10
H	✓ ON		
T	✓	2	

Individual letters make words and words make up sentences (or lines) and lines make up paragraphs. Paragraphs make up columns of text and columns of text make up pages. When handling type we are dealing with designed letterforms. Only very few experienced typographers take the time and considerable effort to design new typefaces. *Though over the*

Column 3 — User Defined H&J

	Min	Opt	Max
W	85	100	250
C	0	0	0
H	OFF		
T	0		

Individual letters make words and words make up sentences (or lines) and lines make up paragraphs. Paragraphs make up columns of text and columns of text make up pages. When handling type we are dealing with designed letterforms. Only very few experienced typographers take the time and considerable effort to design new typefaces. Though over the years thousands of typefaces have been designed, so the choice for us is enormous. When we look at individual typefaces we can see that they each have their own particular characteristics and "personality." We may *examine each of the 26 letters of the alphabet and can see that their form and structure has a special relationship with the background on which it sits. Not only does the shape of the*

Understanding texts and their one another is

letterforms and interaction with very important

The inherent clarity of Gill Sans makes it a good font for typographical experimentation like this negative leading.

COND AaBbCcDdEeFfGgHhIiJjKkLlMmNnOoPpQqRrSsTt
UuVvWwXxYyZz1234567890;:'"!?@ #£$&*()+=

BOLD COND **AaBbCcDdEeFfGgHhIiJjKkLlMmNnOoPpQqRrSsTt
UuVvWwXxYyZz1234567890;:'"!?@#£$&*()+=**

ULTRA BOLD COND **AaBbCcDdEeFfGgHhIiJjKkLlMmNnOoPpQqRrSsTt
UuVvWwXxYyZz1234567890;:'"!?@#£$&*()+=**

*qrstuvwxyz*PAGES122-123

SANS SERIF

Interstate

The prolific American type designer Tobias Frere-Jones designed Interstate in 1993 for The Font Bureau Inc. of Boston, Massachusetts. Based on the American government signage alphabet used on the U.S. highways, Highway Gothic, Interstate is a clean, highly legible design that keeps the familiar quirkiness of its roots but is optically correct. Frere-Jones finds his inspiration from a wide range of sources outside typography. Although a type designer of the digital age, he has the greatest respect for traditional type and printing and the challenges they presented.

The Interstate family has a strong industrial and utilitarian feel and has recently been extended by Frere-Jones, with the assistance of Cyrus Highsmith, to include italics and extreme weights—an indication of its continuing popularity.

TYPE CHARACTERISTICS

The lowercase "e" has an oblique terminal. The two-story lowercase "a" has a small bowl, the lowercase "e" has a low horizontal stroke, and the single-story lowercase "g" has a short tail with a minimal curve. Most of the vertical strokes have an angled finish—particularly noticeable in the Black and Ultra Black weights.

The current Interstate family includes: Hairline, Thin, Extra Light, Light, Regular, Bold, Black, and Ultra Black in roman, italic, condensed, and compressed versions, not all of which can be shown here.

ABCDEFGHIJKLMNOPQRSTUVWXYZ
abcdefghijklmnopqrstuvwxyz1234567890;:'"!?@#£$&*()+=

Quark Default H&J

	Min	Opt	Max
W 🖉	85	110	250
C 🖉	0	0	4
H ✓	ON		
T	0		

Individual letters make words and words make up sentences (or lines) and lines make up paragraphs. Paragraphs make up columns of text and columns of text make up pages. When handling type we are dealing with designed letterforms. Only very few experienced typographers take the time and considerable effort to design new typefaces. Though over the years thousands of typefaces have been designed, so the choice for us is enormous. When we look at individual typefaces we can see that they each have *their own particular characteristics and "personality." We may examine each of the 26 letters of the alphabet and can see that their form and structure has a*

User Defined H&J

	Min	Opt	Max
W 🖉	70	90	110
C 🖉	–6	–3	0
H ✓	ON		
T ✓	3		

Individual letters make words and words make up sentences (or lines) and lines make up paragraphs. Paragraphs make up columns of text and columns of text make up pages. When handling type we are dealing with designed letterforms. Only very few experienced typographers take the time and considerable effort to design new typefaces. Though over the years thousands of typefaces have been designed, so the choice for us is enormous. When we look at individual typefaces we can see that they each have their own *particular characteristics and "personality." We may examine each of the 26 letters of the alphabet and can see that their form and structure has a special rela-*

User Defined H&J

	Min	Opt	Max
W 🖉	70	90	110
C 🖉	–6	–3	0
H ✓	ON		
T	0		

Individual letters make words and words make up sentences (or lines) and lines make up paragraphs. Paragraphs make up columns of text and columns of text make up pages. When handling type we are dealing with designed letterforms. Only very few experienced typographers take the time and considerable effort to design new typefaces. Though over the years thousands of typefaces have been designed, so the choice for us is enormous. When we look at individual typefaces we can see that they each have their own particular characteristics and "personality." We *may examine each of the 26 letters of the alphabet and can see that their form and structure has a special relationship with the background on which it sits. Not only*

Quark Default H&J

	Min	Opt	Max
W	85	110	250
C	0	0	4
H ✓	ON		
T	0		

Individual letters make words and words make up sentences (or lines) and lines make up paragraphs. Paragraphs make up columns of text and columns of text make up pages. When handling type we are dealing with designed letterforms. Only *very few experienced typographers take the time and considerable effort to*

Quark Default H&J

	Min	Opt	Max
W	85	110	250
C	0	0	4
H	OFF		
T	0		

Individual letters make words and words make up sentences (or lines) and lines make up paragraphs. Paragraphs make up columns of text and columns of text make up pages. When handling type we are dealing with designed letterforms. Only *very few experienced typographers take the time and considerable effort to*

Quark Default H&J

	Min	Opt	Max
W	85	110	250
C	0	0	4
H ✓	ON		
T	0		

Individual letters make words and words make up sentences (or lines) and lines make up paragraphs. Paragraphs make up columns of text and columns of text make up pages. When handling type we are dealing with designed letterforms. Only *very few experienced typographers take the time and considerable effort to*

Quark Default H&J

	Min	Opt	Max
W	85	110	250
C	0	0	4
H ✓	ON	HZ 14mm	
T	0		

Individual letters make words and words make up sentences (or lines) and lines make up paragraphs. Paragraphs make up columns of text and columns of text make up pages. When handling type we are dealing with designed letterforms. Only *very few experienced typographers take the time and considerable effort to*

User Defined H&J

	Min	Opt	Max
W	75	90	110
C	-3	-3	3
H ✓	ON		
T	0		

Individual letters make words and words make up sentences (or lines) and lines make up paragraphs. Paragraphs make up columns of text and columns of text make up pages. When handling type we are dealing with designed letterforms. Only very few experienced typographers take the time and considerable effort to design new typefaces. *Though over*

User Defined H&J

	Min	Opt	Max
W	60	90	90
C	-5	0	10
H ✓	ON		
T ✓	2		

Individual letters make words and words make up sentences (or lines) and lines make up paragraphs. Paragraphs make up columns of text and columns of text make up pages. When handling type we are dealing with designed letterforms. Only very few experienced typographers *take the time and considerable effort to design new typefaces. Though over the*

User Defined H&J

	Min	Opt	Max
W	85	100	250
C	0	0	0
H	OFF		
T	0		

Individual letters make words and words make up sentences (or lines) and lines make up paragraphs. Paragraphs make up columns of text and columns of text make up pages. When handling type we are dealing with designed letterforms. Only very few experienced typographers take the time and considerable effort to design new typefaces. Though over the years thousands of typefaces have been designed, so the choice for us is enormous. When we look at individual typefaces we can see that they each have their own particular characteristics and "personality." We may examine *each of the 26 letters of the alphabet and can see that their form and structure has a special relationship with the background on which it sits. Not only does the shape of the*

User Defined H&J

	Min	Opt	Max
W	70	100	120
C	-3	0	5
H	OFF		
T	0		

Individual letters make words and words make up sentences (or lines) and lines make up paragraphs. Paragraphs make up columns of text and columns of text make up pages. When handling type we are dealing with designed letterforms. Only very few experienced typographers take the time and considerable effort to design new typefaces. Though over the years thousands of typefaces have been designed, so the choice for us is enormous. When we look at individual typefaces we can see that they each have their own particular *characteristics and "personality." We may examine each of the 26 letters of the alphabet and can see that their form and structure has a special relationship with the*

HAIRLINE AaBbCcDdEeFfGgHhIiJjKkLlMmNnOoPpQqRrSsTt UuVvWwXxYyZz1234567890;:'"!?@#£$&*()+=

LIGHT AaBbCcDdEeFfGgHhIiJjKkLlMmNnOoPpQqRrSsTt UuVvWwXxYyZz1234567890;:'"!?@#£$&*()+=

BOLD **AaBbCcDdEeFfGgHhIiJjKkLlMmNnOoPpQqRrSsTt UuVvWwXxYyZz1234567890;:'"!?@#£$&*()+=**

BLACK **AaBbCcDdEeFfGgHhIiJjKkLlMmNnOoPpQqRrSsTt UuVvWwXxYyZz1234567890;:'"!?@#£$&*()+=**

LIGHT COND AaBbCcDdEeFfGgHhIiJjKkLlMmNnOoPpQqRrSsTt UuVvWwXxYyZz1234567890;:'"!?@#£$&*()+=

LIGHT COMP AaBbCcDdEeFfGgHhIiJjKkLlMmNnOoPpQqRrSsTt UuVvWwXxYyZz1234567890;:'"!?@#£$&*()+=

COND AaBbCcDdEeFfGgHhIiJjKkLlMmNnOoPpQqRrSsTt UuVvWwXxYyZz1234567890;:'"!?@#£$&*()+=

COMP AaBbCcDdEeFfGgHhIiJjKkLlMmNnOoPpQqRrSsTt UuVvWwXxYyZz1234567890;:'"!?@#£$&*()+=

SANS SERIF

FF Meta

Designed by Erik Spiekermann in 1991, at a point when technological advances in communications, desktop publishing, and "instant print" presented new challenges to traditional typographic standards, Meta captures the spirit of the information technology age. It has a neutral, but distinctive, functional character that Spiekermann describes as having a "rugged charm that owes a lot to the detailed requirements of small type on bad paper." Meta, often referred to as "the Helvetica of the 1990s," is a very adaptable face used for the innovative to the informative and functional. It has proven, excellent legibility across a wide range of contexts, including screen and print. The design of Meta grew out of a design concept for the German postal system of a corporate typeface to replace the long established Helvetica. Fortunately for today's typographic market, it was never implemented. Instead the concept became the basis of an in-house typeface for Spiekermann's design company, MetaDesign, hence its name. The Meta font then became integral to the house style of the FontShop International catalog. Subsequently, it was licensed to FontShop and issued as a widely available FontFont exclusive (FontShop's own label), and quickly became very popular. Meta is one of FontShop's most successful and popular typefaces.

TYPE CHARACTERISTICS

Meta is a contemporary Humanist sans serif with varying stroke widths and angled stroke endings, although the legs of the capitals "K," "R," and "X" sit horizontally on the baseline. The lowercase two-story "g" has an open and almost rectangular loop.

Meta Plus, the latest version of the Meta font, currently has an extensive family of some 39 different weights and styles: Meta Plus Normal, Book, Medium, Bold, Black, and so on.

Unusually, nonlining numerals are standard. Lining numerals are available in alternative "LF" fonts.

ABCDEFGHIJKLMNOPQRSTUVWXYZ
abcdefghijklmnopqrstuvwxyz1234567890;:'"'!?@#£$&*()+=
ABCDEFGHIJKLMNOPQRSTUVWXYZ1234567890

Quark Default H&J

	Min	Opt	Max
W	85	110	250
C	0	0	4
H	ON		
T	0		

Individual letters make words and words make up sentences (or lines) and lines make up paragraphs. Paragraphs make up columns of text and columns of text make up pages. When handling type we are dealing with designed letterforms. Only very few experienced typographers take the time and considerable effort to design new typefaces. Though over the years thousands of typefaces have been designed, so the choice for us is enormous. When we look at individual typefaces we can see that they each have their own particular characteristics and "personality." We may *examine each of the 26 letters of the alphabet and can see that their form and structure has a special relationship with the background on which it sits. Not only does the shape of*

User Defined H&J

	Min	Opt	Max
W	70	100	120
C	-3	0	5
H	ON		
T	3		

Individual letters make words and words make up sentences (or lines) and lines make up paragraphs. Paragraphs make up columns of text and columns of text make up pages. When handling type we are dealing with designed letterforms. Only very few experienced typographers take the time and considerable effort to design new typefaces. Though over the years thousands of typefaces have been designed, so the choice for us is enormous. When we look at individual typefaces we can see that they each have their own particular characteristics and "personality." We may *examine each of the 26 letters of the alphabet and can see that their form and structure has a special relationship with the background on which it sits. Not only does*

User Defined H&J

	Min	Opt	Max
W	70	100	120
C	-1	0	3
H	ON		
T	0		

Individual letters make words and words make up sentences (or lines) and lines make up paragraphs. Paragraphs make up columns of text and columns of text make up pages. When handling type we are dealing with designed letterforms. Only very few experienced typographers take the time and considerable effort to design new typefaces. Though over the years thousands of typefaces have been designed, so the choice for us is enormous. When we look at individual typefaces we can see that they each have their own particular characteristics and "personality." We may examine each of *the 26 letters of the alphabet and can see that their form and structure has a special relationship with the background on which it sits. Not only does the shape of the letter itself*

Quark Default H&J

	Min	Opt	Max
W	85	110	250
C	0	0	4
H ✓	ON		
T	0		

Individual letters make words and words make up sentences (or lines) and lines make up paragraphs. Paragraphs make up columns of text and columns of text make up pages. When handling type we are dealing with designed letterforms. Only very few experienced typographers take *the time and considerable effort to design new typefaces. Though over the years thou-*

Quark Default H&J

	Min	Opt	Max
W	85	110	250
C	0	0	4
H	OFF		
T	0		

Individual letters make words and words make up sentences (or lines) and lines make up paragraphs. Paragraphs make up columns of text and columns of text make up pages. When handling type we are dealing with designed letterforms. Only very few experienced typographers take *the time and considerable effort to design new typefaces. Though over the years*

Quark Default H&J

	Min	Opt	Max
W	85	110	250
C	0	0	4
H ✓	ON		
T	0		

Individual letters make words and words make up sentences (or lines) and lines make up paragraphs. Paragraphs make up columns of text and columns of text make up pages. When handling type we are dealing with designed letterforms. Only very few experienced typographers *take the time and considerable effort to design new typefaces. Though over the*

Quark Default H&J

	Min	Opt	Max
W	85	110	250
C	0	0	4
H ✓	ON HZ 14mm		
T	0		

Individual letters make words and words make up sentences (or lines) and lines make up paragraphs. Paragraphs make up columns of text and columns of text make up pages. When handling type we are dealing with designed letterforms. Only very few experienced typographers *take the time and consider-able effort to design new typefaces. Though over the*

User Defined H&J

	Min	Opt	Max
W	85	100	120
C	-3	0	3
H ✓	ON		
T	0		

Individual letters make words and words make up sentences (or lines) and lines make up paragraphs. Paragraphs make up columns of text and columns of text make up pages. When handling type we are dealing with designed letterforms. Only very few experienced typographers take the time and consid-*erable effort to design new typefaces. Though over the years thousands of typefaces*

User Defined H&J

	Min	Opt	Max
W	75	90	110
C	-3	-3	3
H ✓	ON		
T ✓	2		

Individual letters make words and words make up sentences (or lines) and lines make up paragraphs. Paragraphs make up columns of text and columns of text make up pages. When handling type we are dealing with designed letterforms. Only very few experienced typographers take the time and *considerable effort to design new typefaces. Though over the years thousands of typefaces*

User Defined H&J

	Min	Opt	Max
W	85	100	250
C	0	0	0
H	OFF		
T	0		

Individual letters make words and words make up sentences (or lines) and lines make up paragraphs. Paragraphs make up columns of text and columns of text make up pages. When handling type we are dealing with designed letterforms. Only very few experienced typographers take the time and considerable effort to design new typefaces. Though over the years thousands of typefaces have been designed, so the choice for us is enormous. When we look at individual typefaces we can see that they each have their own particular characteristics and "personality." We may examine each of the 26 letters of the alphabet and can see that their form and *structure has a special relationship with the background on which it sits. Not only does the shape of the letter itself give it form, but the space around it accents its individuality. When two or more*

User Defined H&J

	Min	Opt	Max
W	70	100	120
C	-3	0	5
H	OFF		
T	0		

Individual letters make words and words make up sentences (or lines) and lines make up paragraphs. Paragraphs make up columns of text and columns of text make up pages. When handling type we are dealing with designed letterforms. Only very few experienced typographers take the time and considerable effort to design new typefaces. Though over the years thousands of typefaces have been designed, so the choice for us is enormous. When we look at individual typefaces we can see that they each have their own particular characteristics and "personality." We may examine each of *the 26 letters of the alphabet and can see that their form and structure has a special relationship with the background on which it sits. Not only does the shape of the letter itself*

ITALIC	*AaBbCcDdEeFfGgHhIiJjKkLlMmNnOoPpQqRrSsTt UuVvWwXxYyZz1234567890;:'"!?@#£$&*()+=*
BOOK	AaBbCcDdEeFfGgHhIiJjKkLlMmNnOoPpQqRrSsTt UuVvWwXxYyZz1234567890;:'"!?@#£$&*()+=
BOOK ITALIC	*AaBbCcDdEeFfGgHhIiJjKkLlMmNnOoPpQqRrSsTt UuVvWwXxYyZz1234567890;:' "!?@#£$&*()+=*
MEDIUM	**AaBbCcDdEeFfGgHhIiJjKkLlMmNnOoPpQqRrSsTt UuVvWwXxYyZz1234567890;:'"!?@#£$&*()+=**
MEDIUM ITALIC	*AaBbCcDdEeFfGgHhIiJjKkLlMmNnOoPpQqRrSsTt UuVvWwXxYyZz1234567890;:' "!?@#£$&*()+=*

FF Meta Bold

Meta

Setting tips

Meta can be used successfully at very small sizes. Its open, clean design is most legible when not tightly tracked.

Meta
Meta
Meta
Meta

Normal spacing between characters remains similar from weight to weight.

1650

Rene Descartes, philosopher, dies

1651

Cromwell defeats King Charles II

1652

Massachusetts declares independence

Meta nonlining numerals

qQnN

The evenness of weight found in the small caps allows for interesting mixing and mingling of small caps with lowercase letters.

AaBbCc

ABCDEFGHIJKLMNOPQRSTUVWXYZ
abcdefghijklmnopqrstuvwxyz1234567890;:'"!?@#£$&*()+=
ABCDEFGHIJKLMNOPQRSTUVWXYZ1234567890;:'"!?@#£$&*()+=

Quark Default H&J

	Min	Opt	Max
W ✐	85	110	250
C ✐	0	0	4
H ✓	ON		
T	0		

Individual letters make words and words make up sentences (or lines) and lines make up paragraphs. Paragraphs make up columns of text and columns of text make up pages. When handling type we are dealing with designed letterforms. Only very few experienced typographers take the time and considerable effort to design new typefaces. Though over the years thousands of typefaces have been designed, so the choice for us is enormous. When we look at individual typefaces we can see that they each have their own particular characteristics and "per-*sonality." We may examine each of the 26 letters of the alphabet and can see that their form and structure has a special relationship with the background on which it sits.*

User Defined H&J

	Min	Opt	Max
W ✐	90	100	120
C ✐	0	3	10
H ✓	ON		
T ✓	3		

Individual letters make words and words make up sentences (or lines) and lines make up paragraphs. Paragraphs make up columns of text and columns of text make up pages. When handling type we are dealing with designed letterforms. Only very few experienced typographers take the time and considerable effort to design new typefaces. Though over the years thousands of typefaces have been designed, so the choice for us is enormous. When we look at individual typefaces we can see that they each have their own par-*ticular characteristics and "personality." We may examine each of the 26 letters of the alphabet and can see that their form and structure has a special relationship*

User Defined H&J

	Min	Opt	Max
W ✐	70	100	120
C ✐	–3	0	5
H ✓	ON		
T	0		

User Defined H&J

	Min	Opt	Max
W ✐	70	90	110
C ✐	–6	–3	0
H	OFF		
T	0		

Individual letters make words and words make up sentences (or lines) and lines make up paragraphs. Paragraphs make up columns of text and columns of text make up pages. When handling type we are dealing with designed letterforms. Only very few experienced typographers take the time and considerable effort to design new typefaces. Though over

Individual letters make words and words make up sentences (or lines) and lines make up paragraphs. Paragraphs make up columns of text and columns of text make up pages. When handling type we are dealing with designed letterforms. Only very few experienced typographers take the time and considerable effort to design new typefaces. Though over the

Quark Default H&J

	Min	Opt	Max
W	85	110	250
C	0	0	4
H ✓	ON		
T	0		

Individual letters make words and words make up sentences (or lines) and lines make up paragraphs. Paragraphs make up columns of text and columns of text make up pages. When handling type we are dealing with designed letterforms. Only very few experienced typographers *take the time and considerable effort to design new typefaces. Though over the*

Quark Default H&J

	Min	Opt	Max
W	85	110	250
C	0	0	4
H	OFF		
T	0		

Individual letters make words and words make up sentences (or lines) and lines make up paragraphs. Paragraphs make up columns of text and columns of text make up pages. When handling type we are dealing with designed letterforms. Only very few experienced typographers *take the time and considerable effort to design new typefaces. Though over*

Quark Default H&J

	Min	Opt	Max
W	85	110	250
C	0	0	4
H ✓	ON		
T	0		

Individual letters make words and words make up sentences (or lines) and lines make up paragraphs. Paragraphs make up columns of text and columns of text make up pages. When handling type we are dealing with designed letterforms. Only very few experienced typographers *take the time and considerable effort to design new typefaces. Though over the*

Quark Default H&J

	Min	Opt	Max
W	85	110	250
C	0	0	4
H ✓	ON	HZ 14mm	
T	0		

Individual letters make words and words make up sentences (or lines) and lines make up paragraphs. Paragraphs make up columns of text and columns of text make up pages. When handling type we are dealing with designed letterforms. Only very few experienced typographers *take the time and considerable effort to design new typefaces. Though over the*

User Defined H&J

	Min	Opt	Max
W	60	90	110
C	–5	–5	5
H ✓	ON		
T	0		

Individual letters make words and words make up sentences (or lines) and lines make up paragraphs. Paragraphs make up columns of text and columns of text make up pages. When handling type we are dealing with designed letterforms. Only very few experienced typographers *take the time and considerable effort to design new typefaces. Though over the years thousands of typefaces have*

User Defined H&J

	Min	Opt	Max
W	60	90	90
C	–5	0	10
H ✓	ON		
T ✓	2		

Individual letters make words and words make up sentences (or lines) and lines make up paragraphs. Paragraphs make up columns of text and columns of text make up pages. When handling type we are dealing with designed letterforms. Only very few experienced typographers *take the time and considerable effort to design new typefaces. Though over the years thousands of typefaces*

User Defined H&J

	Min	Opt	Max
W	85	100	250
C	0	0	0
H	OFF		
T	0		

Individual letters make words and words make up sentences (or lines) and lines make up paragraphs. Paragraphs make up columns of text and columns of text make up pages. When handling type we are dealing with designed letterforms. Only very few experienced typographers take the time and considerable effort to design new typefaces. Though over the years thousands of typefaces have been designed, so the choice for us is enormous. When we look at individual typefaces we can see that they each have their own particular characteristics and "personality." We may examine each of the 26 letters of the *alphabet and can see that their form and structure has a special relationship with the background on which it sits. Not only does the shape of the letter itself give it form, but the space around it*

Meta Roman showing overall tracking of –10 with letter pairs individually kerned as shown. Gray text indicates 0 tracking and no kerning of letter pairs.

BOLD **AaBbCcDdEeFfGgHhIiJjKkLlMmNnOoPpQqRrSsTt UuVvWwXxYyZz1234567890;:'"!?@#£$&*()+=**

BOLD ITALIC ***AaBbCcDdEeFfGgHhIiJjKkLlMmNnOoPpQqRrSsTt UuVvWwXxYyZz1234567890;:'"!?@#£$&*()+=***

BLACK **AaBbCcDdEeFfGgHhIiJjKkLlMmNnOoPpQqRrSsTt UuVvWwXxYyZz1234567890;:'"!?@#£$&*()+=**

BLACK ITALIC ***AaBbCcDdEeFfGgHhIiJjKkLlMmNnOoPpQqRrSsTt UuVvWwXxYyZz1234567890;:'"!?@#£$&*()+=***

Myriad Multiple Master (MM)

Myriad Multiple Master (together with Minion) was one of the first typeface "families" with new technology integrated into the design concept. Jointly designed in 1991 by Robert Slimbach and Carol Twombly, Myriad MM lets the user determine the exact size, weight, and width of character. By altering one or more of the design axes, the user "creates" a customized, virtually unlimited type family of consistent proportions, known as "instances." Multiple Masters can, theoretically, incorporate up to five axes, which may include an optical size axis that ensures good readability across a wide range of type sizes—a contemporary equivalent to the traditional fine-tuning applied to hand-cut metal type. In practice, most Multiple Master fonts, including Myriad, work with just two axes—weight and width.

TYPE CHARACTERISTICS

Myriad has a strong, rounded look with some extremely subtle changes in stroke weight. The capital "S" has a distinctively open form and width. The lowercase letters have an angled finish to all the curved strokes.

DESIGN AXIS

WEIGHT — aaaaaaaaaa LIGHT TO BLACK

WIDTH — aaaaaaaa CONDENSED TO EXTRA EXTENDED

STYLE — aaaaaaaaaa WEDGE SERIF TO SLAB SERIF

OPTICAL SIZE — aaaaaaaaaa 6-PT TO 72-PT (SCALED TO SAME SIZE)

DYNAMIC RANGE

ABCDEFGHIJKLMNOPQRSTUVWXYZ
abcdefghijklmnopqrstuvwxyz1234567890;:'"!?@#£$&*()+=

Quark Default H&J

	Min	Opt	Max
W	85	110	250
C	0	0	4
H	ON		
T	0		

Individual letters make words and words make up sentences (or lines) and lines make up paragraphs. Paragraphs make up columns of text and columns of text make up pages. When handling type we are dealing with designed letterforms. Only very few experienced typographers take the time and considerable effort to design new typefaces. Though over the years thousands of typefaces have been designed, so the choice for us is enormous. When we look at individual typefaces we can see that they each have their own particular characteristics and "personality." We may examine each of the 26 letters of the alphabet and can see *that their form and structure has a special relationship with the background on which it sits. Not only does the shape of the letter itself give it form, but the space around it accents* its

User Defined H&J

	Min	Opt	Max
W	70	100	105
C	0	0	3
H	ON		
T	3		

Individual letters make words and words make up sentences (or lines) and lines make up paragraphs. Paragraphs make up columns of text and columns of text make up pages. When handling type we are dealing with designed letterforms. Only very few experienced typographers take the time and considerable effort to design new typefaces. Though over the years thousands of typefaces have been designed, so the choice for us is enormous. When we look at individual typefaces we can see that they each have their own particular characteristics and "personality." We *may examine each of the 26 letters of the alphabet and can see that their form and structure has a special relationship with the background on which it sits. Not only does the shape*

User Defined H&J

	Min	Opt	Max
W	70	100	120
C	0	0	6
H	ON		
T	0		

Individual letters make words and words make up sentences (or lines) and lines make up paragraphs. Paragraphs make up columns of text and columns of text make up pages. When handling type we are dealing with designed letterforms. Only very few experienced typographers take the time and considerable effort to design new typefaces. Though over the years thousands of typefaces have been designed, so the choice for us is enormous. When we look at individual typefaces we can see that they each have their own particular characteristics and "personality." We may *examine each of the 26 letters of the alphabet and can see that their form and structure has a special relationship with the background on which it sits. Not only does the shape of the let-*

	Min	Opt	Max
W	85	110	250
C	0	0	4
H	✓ ON		
T	0		

Individual letters make words and words make up sentences (or lines) and lines make up paragraphs. Paragraphs make up columns of text and columns of text make up pages. When handling type we are dealing with designed letterforms. Only very few experienced typographers take *the time and considerable effort to design new typefaces. Though over the years thousands of type-*

	Min	Opt	Max
W	85	110	250
C	0	0	4
H	OFF		
T	0		

Individual letters make words and words make up sentences (or lines) and lines make up paragraphs. Paragraphs make up columns of text and columns of text make up pages. When handling type we are dealing with designed letterforms. Only very few experienced typographers take *the time and considerable effort to design new typefaces. Though over the years thousands of*

	Min	Opt	Max
W	85	110	250
C	0	0	4
H	✓ ON		
T	0		

Individual letters make words and words make up sentences (or lines) and lines make up paragraphs. Paragraphs make up columns of text and columns of text make up pages. When handling type we are dealing with designed letterforms. Only very few experienced typographers *take the time and considerable effort to design new typefaces. Though over the years thou-*

	Min	Opt	Max
W	85	110	250
C	0	0	4
H	✓ ON HZ 14mm		
T	0		

Individual letters make words and words make up sentences (or lines) and lines make up paragraphs. Paragraphs make up columns of text and columns of text make up pages. When handling type we are dealing with designed letterforms. Only very few experienced typographers *take the time and considerable effort to design new typefaces. Though over the years*

	Min	Opt	Max
W	70	90	115
C	-3	0	5
H	✓ ON		
T	0		

Individual letters make words and words make up sentences (or lines) and lines make up paragraphs. Paragraphs make up columns of text and columns of text make up pages. When handling type we are dealing with designed letterforms. Only very few experienced typographers take the time and considerable *effort to design new typefaces. Though over the years thousands of typefaces have been designed,*

	Min	Opt	Max
W	60	90	90
C	-5	0	10
H	✓ ON		
T	✓ 2		

Individual letters make words and words make up sentences (or lines) and lines make up paragraphs. Paragraphs make up columns of text and columns of text make up pages. When handling type we are dealing with designed letterforms. Only very few experienced typographers take the time and con*siderable effort to design new typefaces. Though over the years thousands of typefaces have been*

	Min	Opt	Max
W	85	100	250
C	0	0	0
H	OFF		
T	0		

Individual letters make words and words make up sentences (or lines) and lines make up paragraphs. Paragraphs make up columns of text and columns of text make up pages. When handling type we are dealing with designed letterforms. Only very few experienced typographers take the time and considerable effort to design new typefaces. Though over the years thousands of typefaces have been designed, so the choice for us is enormous. When we look at individual typefaces we can see that they each have their own particular characteristics and "personality." We may examine each of the 26 letters of the alphabet and can see that their form and *structure has a special relationship with the background on which it sits. Not only does the shape of the letter itself give it form, but the space around it accents its individuality. When two or more letters are*

	Min	Opt	Max
W	70	100	120
C	-1	0	3
H	OFF		
T	0		

Individual letters make words and words make up sentences (or lines) and lines make up paragraphs. Paragraphs make up columns of text and columns of text make up pages. When handling type we are dealing with designed letterforms. Only very few experienced typographers take the time and considerable effort to design new typefaces. Though over the years thousands of typefaces have been designed, so the choice for us is enormous. When we look at individual typefaces we can see that they each have their own particular characteristics and "personality." We may examine each of *the 26 letters of the alphabet and can see that their form and structure has a special relationship with the background on which it sits. Not only does the shape of the letter itself give it*

ITALIC *AaBbCcDdEeFfGgHhIiJjKkLlMmNnOoPpQqRrSsTt UuVvWwXxYyZz1234567890;:'"!?@#£$&*()+=*

BOLD **AaBbCcDdEeFfGgHhIiJjKkLlMmNnOoPpQqRrSsTt UuVvWwXxYyZz1234567890;:'"!?@#£$&*()+=**

BOLD ITALIC ***AaBbCcDdEeFfGgHhIiJjKkLlMmNnOoPpQqRrSsTt UuVvWwXxYyZz1234567890;:'"!?@#£$&*()+=***

Neue Helvetica

Neue Helvetica grew out of its predecessor, Helvetica, originally designed as Neue Haas Grotesk in 1957 by the Swiss type designer Max Miedinger. With its functional style, Helvetica became very popular in the 1960s and 1970s. Different versions were introduced, giving rise to confusion with inevitable design inconsistencies.

In 1983, Linotype redrew the entire Helvetica family as an integrated concept. Each of these new weights and styles carried a standard numeric reference, from 25 (ultra light) to 95 (ultra heavy), to avoid confusion.

Neue Helvetica has a coherence, clarity, and visual efficiency that, together with its international feel, makes it highly usable and easily readable. The Helvetica family is strongly associated with the clean, ordered Swiss design of the 1960s. Together with Times (New) Roman, Helvetica remains one of the most widely used typefaces, not least due to its adoption as a system font for Macs and as a default font for QuarkXPress.

TYPE CHARACTERISTICS

Neue Helvetica has well-proportioned characters with a large x-height and an even, rhythmic feel. The large family of weights and widths—Linotype lists some 115 fonts—allows for considerable choice in the coherent visual structuring of hierarchical information. This versatile range lends itself equally well to textural color and contrast in text matter, as well as display and signage contexts. The simplicity of its design, free from emotion or decoration, makes it highly suitable for the presentation of information.

Helvetica has few unique capital letterforms apart from the "G" with a long serif on a high bar and a slightly tapered spur, the double junction on the "K" (which also appears in the lowercase "k"), and the wide-apart junction and slightly curled leg of the "R." The capital "Q" has a no-nonsense, but somewhat aggressive, tail in the form of a diagonal cross-stroke. In the lowercase letters, the two-story "a," the narrow "f" and "t," together with the fat "s," are typical of Helvetica. All of the curved strokes end horizontally.

ABCDEFGHIJKLMNOPQRSTUVWXYZ
abcdefghijklmnopqrstuvwxyz1234567890;:'"!?@#£$&*()+=

Quark Default H&J

	Min	Opt	Max
W	85	110	250
C	0	0	4
H ✓	ON		
T	0		

Individual letters make words and words make up sentences (or lines) and lines make up paragraphs. Paragraphs make up columns of text and columns of text make up pages. When handling type we are dealing with designed letterforms. Only very few experienced typographers take the time and considerable effort to design new typefaces. Though over the years thousands of typefaces have been designed, so the choice for us is enormous. When we look at individual typefaces we can see that they each have their own *particular characteristics and "personality." We may examine each of the 26 letters of the alphabet and can see that their form and structure has a special relation-*

User Defined H&J

	Min	Opt	Max
W	80	90	150
C	0	3	5
H ✓	ON		
T ✓	3		

Individual letters make words and words make up sentences (or lines) and lines make up paragraphs. Paragraphs make up columns of text and columns of text make up pages. When handling type we are dealing with designed letterforms. Only very few experienced typographers take the time and considerable effort to design new typefaces. Though over the years thousands of typefaces have been designed, so the choice for us is enormous. When we look at individual typefaces we can see that they each have their own *particular characteristics and "personality." We may examine each of the 26 letters of the alphabet and can see that their form and structure has a special rela-*

User Defined H&J

	Min	Opt	Max
W	70	90	110
C	−6	−3	0
H ✓	ON		
T	0		

Individual letters make words and words make up sentences (or lines) and lines make up paragraphs. Paragraphs make up columns of text and columns of text make up pages. When handling type we are dealing with designed letterforms. Only very few experienced typographers take the time and considerable effort to design new typefaces. Though over the years thousands of typefaces have been designed, so the choice for us is enormous. When we look at individual typefaces we can see that they each have their own particular characteristics and "personality." We may *examine each of the 26 letters of the alphabet and can see that their form and structure has a special relationship with the background on which it sits. Not only does the shape*

Column 1

Quark Default H&J

	Min	Opt	Max
W	85	110	250
C	0	0	4
H ✓	ON		
T	0		

Individual letters make words and words make up sentences (or lines) and lines make up paragraphs. Paragraphs make up columns of text and columns of text make up pages. When handling type we are dealing with designed letterforms. Only very few experienced *typographers take the time and considerable effort to design new typefaces.*

User Defined H&J

	Min	Opt	Max
W	60	90	100
C	-3	0	5
H ✓	ON		
T	0		

Individual letters make words and words make up sentences (or lines) and lines make up paragraphs. Paragraphs make up columns of text and columns of text make up pages. When handling type we are dealing with designed letterforms. Only very few experienced typographers take the *time and considerable effort to design new typefaces. Though over the years thousands of*

User Defined H&J

	Min	Opt	Max
W	60	100	120
C	-3	0	5
H	OFF		
T	0		

Individual letters make words and words make up sentences (or lines) and lines make up paragraphs. Paragraphs make up columns of text and columns of text make up pages. When handling type we are dealing with designed letterforms. Only very few experienced typographers take the time and considerable effort to design new typefaces. Though over the years thousands of typefaces have been designed, so the choice for us is enormous. When we look at individual typefaces we can see that they each have their own particular characteristics *and "personality." We may examine each of the 26 letters of the alphabet and can see that their form and structure has a special relationship with the background on which*

Column 2

Quark Default H&J

	Min	Opt	Max
W	85	110	250
C	0	0	4
H	OFF		
T	0		

Individual letters make words and words make up sentences (or lines) and lines make up paragraphs. Paragraphs make up columns of text and columns of text make up pages. When handling type we are dealing with designed letterforms. Only very few experienced *typographers take the time and considerable effort to design new typefaces.*

User Defined H&J

	Min	Opt	Max
W	70	90	115
C	-3	0	5
H ✓	ON		
T ✓	2		

Individual letters make words and words make up sentences (or lines) and lines make up paragraphs. Paragraphs make up columns of text and columns of text make up pages. When handling type we are dealing with designed letterforms. Only very few experienced typographers take the *time and considerable effort to design new typefaces. Though over*

Column 3

Quark Default H&J

	Min	Opt	Max
W	85	110	250
C	0	0	4
H ✓	ON		
T	0		

Individual letters make words and words make up sentences (or lines) and lines make up paragraphs. Paragraphs make up columns of text and columns of text make up pages. When handling type we are dealing with designed letterforms. Only *very few experienced typographers take the time and considerable effort to design*

User Defined H&J

	Min	Opt	Max
W	85	100	250
C	0	0	0
H	OFF		
T	0		

Individual letters make words and words make up sentences (or lines) and lines make up paragraphs. Paragraphs make up columns of text and columns of text make up pages. When handling type we are dealing with designed letterforms. Only very few experienced typographers take the time and considerable effort to design new typefaces. Though over the years thousands of typefaces have been designed, so the choice for us is enormous. When we look at individual typefaces we can see that they each have their own particular characteristics and "personality." We may examine each of the *26 letters of the alphabet and can see that their form and structure has a special relationship with the background on which it sits. Not only does the shape of the letter itself give it*

Column 4

Quark Default H&J

	Min	Opt	Max
W	85	110	250
C	0	0	4
H ✓	ON	HZ 14mm	
T	0		

Individual letters make words and words make up sentences (or lines) and lines make up paragraphs. Paragraphs make up columns of text and columns of text make up pages. When handling type we are dealing with designed letterforms. Only *very few experienced typographers take the time and considerable effort to*

Typeface specimens

Weight	Specimen
25 ULTRA LIGHT	AaBbCcDdEeFfGgHhIiJjKkLlMmNnOoPpQqRrSsTt UuVvWwXxYyZz1234567890;:'"!?@#£$&*()+=
26 ULTRA LIGHT ITALIC	*AaBbCcDdEeFfGgHhIiJjKkLlMmNnOoPpQqRrSsTt UuVvWwXxYyZz1234567890;:'"!?@#£$&*()+=*
35 THIN	AaBbCcDdEeFfGgHhIiJjKkLlMmNnOoPpQqRrSsTt UuVvWwXxYyZz1234567890;:'"!?@#£$&*()+=
36 THIN ITALIC	*AaBbCcDdEeFfGgHhIiJjKkLlMmNnOoPpQqRrSsTt UuVvWwXxYyZz1234567890;:'"!?@#£$&*()+=*
45 LIGHT	AaBbCcDdEeFfGgHhIiJjKkLlMmNnOoPpQqRrSsTt UuVvWwXxYyZz1234567890;:'"!?@#£$&*()+=
46 LIGHT ITALIC	*AaBbCcDdEeFfGgHhIiJjKkLlMmNnOoPpQqRrSsTt UuVvWwXxYyZz1234567890;:'"!?@#£$&*()+=*
56 ITALIC	*AaBbCcDdEeFfGgHhIiJjKkLlMmNnOoPpQqRrSsTt UuVvWwXxYyZz1234567890;:'"!?@#£$&*()+=*

Neue Helvetica Condensed·

Neue Helvetica

NORMAL

AAAAAAAA

EXTENDED

AAAAAAAAA

CONDENSED

AAAAAAA

Such a wide range of weights and widths offers opportunities not available to lesser font families.

SQUEEZE

SLIMMING

AaBbCc

ABCDEFGHIJKLMNOPQRSTUVWXYZ
abcdefghijklmnopqrstuvwxyz1234567890;:'"!?@#£$&*()+=

Quark Default H&J			
	Min	Opt	Max
W ✎	85	110	250
C ✎	0	0	4
H ✓	ON		
T	0		

Individual letters make words and words make up sentences (or lines) and lines make up paragraphs. Paragraphs make up columns of text and columns of text make up pages. When handling type we are dealing with designed letterforms. Only very few experienced typographers take the time and considerable effort to design new typefaces. Though over the years thousands of typefaces have been designed, so the choice for us is enormous. When we look at individual typefaces we can see that they each have their own particular characteristics and "personality." We may examine each of the 26 letters of the alphabet and can see that their form and structure has a special relationship *with the background on which it sits. Not only does the shape of the letter itself give it form, but the space around it accents its individuality. When two or more letters are put side by side they create words*

User Defined H&J			
	Min	Opt	Max
W ✎	90	100	100
C ✎	0	0	3
H ✓	ON		
T ✓	3		

Individual letters make words and words make up sentences (or lines) and lines make up paragraphs. Paragraphs make up columns of text and columns of text make up pages. When handling type we are dealing with designed letterforms. Only very few experienced typographers take the time and considerable effort to design new typefaces. Though over the years thousands of typefaces have been designed, so the choice for us is enormous. When we look at individual typefaces we can see that they each have their own particular characteristics and "personality." We may examine each of the 26 letters of the alphabet and can see that their form and structure *has a special relationship with the background on which it sits. Not only does the shape of the letter itself give it form, but the space around it accents its individuality. When two or more letters are put*

User Defined H&J			
	Min	Opt	Max
W ✎	70	90	90
C ✎	3	3	4
H ✓	ON		
T	0		

Individual letters make words and words make up sentences (or lines) and lines make up paragraphs. Paragraphs make up columns of text and columns of text make up pages. When handling type we are dealing with designed letterforms. Only very few experienced typographers take the time and considerable effort to design new typefaces. Though over the years thousands of typefaces have been designed, so the choice for us is enormous. When we look at individual typefaces we can see that they each have their own particular characteristics and "personality." We may examine each of the 26 letters of the alphabet and can see that their form and structure has a special relationship *with the background on which it sits. Not only does the shape of the letter itself give it form, but the space around it accents its individuality. When two or more letters are put side by side they create*

Quark Default H&J	Min	Opt	Max
W	85	110	250
C	0	0	4
H ✓ ON			
T	0		

Individual letters make words and words make up sentences (or lines) and lines make up paragraphs. Paragraphs make up columns of text and columns of text make up pages. When handling type we are dealing with designed letterforms. Only very few experienced typographers take the time and considerable effort to design new typefaces. Though over *the years thousands of typefaces have been designed, so the choice for us is enormous. When we look at*

Quark Default H&J	Min	Opt	Max
W	85	110	250
C	0	0	4
H OFF			
T	0		

Individual letters make words and words make up sentences (or lines) and lines make up paragraphs. Paragraphs make up columns of text and columns of text make up pages. When handling type we are dealing with designed letterforms. Only very few experienced typographers take the time and considerable effort to design new typefaces. Though over *the years thousands of typefaces have been designed, so the choice for us is enormous. When we look at*

Quark Default H&J	Min	Opt	Max
W	85	110	250
C	0	0	4
H ✓ ON			
T	0		

Individual letters make words and words make up sentences (or lines) and lines make up paragraphs. Paragraphs make up columns of text and columns of text make up pages. When handling type we are dealing with designed letterforms. Only very few experienced typographers take the time and considerable effort to design new *typefaces. Though over the years thousands of typefaces have been designed, so the choice for us is*

Quark Default H&J	Min	Opt	Max
W	85	110	250
C	0	0	4
H ✓ ON HZ 14mm			
T	0		

Individual letters make words and words make up sentences (or lines) and lines make up paragraphs. Paragraphs make up columns of text and columns of text make up pages. When handling type we are dealing with designed letterforms. Only very few experienced typographers take the time and considerable effort to design new *typefaces. Though over the years thousands of typefaces have been designed, so the choice for us is*

User Defined H&J	Min	Opt	Max
W	70	90	115
C	-3	0	5
H ✓ ON			
T	0		

Individual letters make words and words make up sentences (or lines) and lines make up paragraphs. Paragraphs make up columns of text and columns of text make up pages. When handling type we are dealing with designed letterforms. Only very few experienced typographers take the time and considerable effort to design new typefaces. Though over the years *thousands of typefaces have been designed, so the choice for us is enormous. When we look at individual type-*

User Defined H&J	Min	Opt	Max
W	85	100	120
C	-3	0	3
H ✓ ON			
T ✓	2		

Individual letters make words and words make up sentences (or lines) and lines make up paragraphs. Paragraphs make up columns of text and columns of text make up pages. When handling type we are dealing with designed letterforms. Only very few experienced typographers take the time and considerable effort to design new typefaces. Though over *the years thousands of typefaces have been designed, so the choice for us is enormous. When we look at*

User Defined H&J	Min	Opt	Max
W	85	100	250
C	0	0	0
H OFF			
T	0		

Individual letters make words and words make up sentences (or lines) and lines make up paragraphs. Paragraphs make up columns of text and columns of text make up pages. When handling type we are dealing with designed letterforms. Only very few experienced typographers take the time and considerable effort to design new typefaces. Though over the years thousands of typefaces have been designed, so the choice for us is enormous. When we look at individual typefaces we can see that they each have their own particular characteristics and "personality." We may examine each of the 26 letters of the alphabet and can see that their form and structure has a special relationship with the background on which it sits. Not only does the shape of the letter itself give it *form, but the space around it accents its individuality. When two or more letters are put side by side they create words which are recognized as complete units, rather than collections of individual shapes. Typeset words create for us a visual*

User Defined H&J	Min	Opt	Max
W	60	100	120
C	-3	0	5
H OFF			
T	0		

Individual letters make words and words make up sentences (or lines) and lines make up paragraphs. Paragraphs make up columns of text and columns of text make up pages. When handling type we are dealing with designed letterforms. Only very few experienced typographers take the time and considerable effort to design new typefaces. Though over the years thousands of typefaces have been designed, so the choice for us is enormous. When we look at individual typefaces we can see that they each have their own particular characteristics and "personality." We may examine each of the 26 letters of the alphabet and can see that their form and structure has a special relationship with the background *on which it sits. Not only does the shape of the letter itself give it form, but the space around it accents its individuality. When two or more letters are put side by side they create words which are recognized as*

47 LIGHT COND
AaBbCcDdEeFfGgHhIiJjKkLlMmNnOoPpQqRrSsTt
UuVvWwXxYyZz1234567890;:'"!?@#£$&*()+=

48 LIGHT COND ITALIC
*AaBbCcDdEeFfGgHhIiJjKkLlMmNnOoPpQqRrSsTt
UuVvWwXxYyZz1234567890;:'"!?@#£$&*()+=*

57 COND
AaBbCcDdEeFfGgHhIiJjKkLlMmNnOoPpQqRrSsTt
UuVvWwXxYyZz1234567890;:'"!?@#£$&*()+=

58 COND ITALIC
*AaBbCcDdEeFfGgHhIiJjKkLlMmNnOoPpQqRrSsTt
UuVvWwXxYyZz1234567890;:'"!?@#£$&*()+=*

67 MEDIUM COND
**AaBbCcDdEeFfGgHhIiJjKkLlMmNnOoPpQqRrSsTt
UuVvWwXxYyZz1234567890;:'"!?@#£$&*()+=**

68 MEDIUM COND ITALIC
***AaBbCcDdEeFfGgHhIiJjKkLlMmNnOoPpQqRrSsTt
UuVvWwXxYyZz1234567890;:'"!?@#£$&*()+=***

Optima

Designed in 1958 by Hermann Zapf, the influential German type designer, this individual typeface design remains one of his most successful and popular typefaces. Optima was originally called New Antiqua by Zapf, but the name was not considered to be marketable and, as the font optically bridges both sans and serif families, the name was changed to Optima, although it can still also be found under the name of Zapf Humanist. Inspired by the proportions of inscriptional Roman capital letters, Optima inventively combines grace with the directness of a sans serif form, resulting in a tranquil design that is easy to read.

TYPE CHARACTERISTICS

Zapf's Optima, cut by the Frankfurt foundry, Stempel, in 1958, has a mix of characteristics. There is an absence of feet at the terminals—a major sans serif feature—yet it has the typographic color typical of seriffed faces. The functional flare in the strokes visually links characters across lines of text and effortlessly leads the eye over the page.

The capital letters are wide, open letters with the exception of the "E," "F," and "L," which are narrower. The capital "M" is slightly splayed and the "S" appears rather top heavy, and carries the typographic signature of most of Zapf's typefaces, a slight forward tilt.

In the lowercase letters, the "a" and "g" are two-story as in seriffed type, the "f" has an angled finish to the top of its stem, the "g" has a horizontal ear that hangs parallel to the baseline, and there is no loop to the "y."

Zapf Humanist, Bitstream's version of Optima, shows the flared strokes best. It makes a large family of different weights ranging from medium to extra black including a demibold. In all versions of Optima there is a slanted roman or oblique "italic."

Apart from at very small sizes, Optima works extremely well both as a text and display face, conveying a modern classicism. It can take generous line spacing and mixes well with other faces.

ABCDEFGHIJKLMNOPQRSTUVWXYZ
abcdefghijklmnopqrstuvwxyz1234567890;:'"!?@#£$&*()+=

Quark Default H&J

	Min	Opt	Max
W	85	110	250
C	0	0	4
H ✓	ON		
T	0		

Individual letters make words and words make up sentences (or lines) and lines make up paragraphs. Paragraphs make up columns of text and columns of text make up pages. When handling type we are dealing with designed letterforms. Only very few experienced typographers take the time and considerable effort to design new typefaces. Though over the years thousands of typefaces have been designed, so the choice for us is enormous. When we look at individual typefaces we can see that they each have their own particular characteristics and "personality." We *may examine each of the 26 letters of the alphabet and can see that their form and structure has a special relationship with the background on which it sits. Not only*

User Defined H&J

	Min	Opt	Max
W	70	90	90
C	3	3	4
H ✓	ON		
T ✓	3		

Individual letters make words and words make up sentences (or lines) and lines make up paragraphs. Paragraphs make up columns of text and columns of text make up pages. When handling type we are dealing with designed letterforms. Only very few experienced typographers take the time and considerable effort to design new typefaces. Though over the years thousands of typefaces have been designed, so the choice for us is enormous. When we look at individual typefaces we can see that they each have their own *particular characteristics and "personality." We may examine each of the 26 letters of the alphabet and can see that their form and structure has a special relation-*

User Defined H&J

	Min	Opt	Max
W	60	100	120
C	–3	0	5
H ✓	ON		
T	0		

Individual letters make words and words make up sentences (or lines) and lines make up paragraphs. Paragraphs make up columns of text and columns of text make up pages. When handling type we are dealing with designed letterforms. Only very few experienced typographers take the time and considerable effort to design new typefaces. Though over the years thousands of typefaces have been designed, so the choice for us is enormous. When we look at individual typefaces we can see that they each have their own particular characteristics and "personality." We may *examine each of the 26 letters of the alphabet and can see that their form and structure has a special relationship with the background on which it sits. Not only does the shape of*

Individual letters make words and words make up sentences (or lines) and lines make up paragraphs. Paragraphs make up columns of text and columns of text make up pages. When handling type we are dealing with designed letterforms. Only very few experienced typographers *take the time and considerable effort to design new typefaces. Though over the*

Individual letters make words and words make up sentences (or lines) and lines make up paragraphs. Paragraphs make up columns of text and columns of text make up pages. When handling type we are dealing with designed letterforms. Only very few experienced typographers *take the time and considerable effort to design new typefaces. Though over*

Individual letters make words and words make up sentences (or lines) and lines make up paragraphs. Paragraphs make up columns of text and columns of text make up pages. When handling type we are dealing with designed letterforms. Only very few experienced *typographers take the time and considerable effort to design new typefaces. Though*

Individual letters make words and words make up sentences (or lines) and lines make up paragraphs. Paragraphs make up columns of text and columns of text make up pages. When handling type we are dealing with designed letterforms. Only very few experienced *typographers take the time and considerable effort to design new typefaces. Though*

Individual letters make words and words make up sentences (or lines) and lines make up paragraphs. Paragraphs make up columns of text and columns of text make up pages. When handling type we are dealing with designed letterforms. Only very few experienced typographers take the time and considerable *effort to design new typefaces. Though over the years thousands of typefaces have been designed,*

Individual letters make words and words make up sentences (or lines) and lines make up paragraphs. Paragraphs make up columns of text and columns of text make up pages. When handling type we are dealing with designed letterforms. Only very few experienced typographers *take the time and considerable effort to design new typefaces. Though over the*

Individual letters make words and words make up sentences (or lines) and lines make up paragraphs. Paragraphs make up columns of text and columns of text make up pages. When handling type we are dealing with designed letterforms. Only very few experienced typographers take the time and considerable effort to design new typefaces. Though over the years thousands of typefaces have been designed, so the choice for us is enormous. When we look at individual typefaces we can see that they each have their own particular characteristics and "personality." We may examine each of the 26 letters of the *alphabet and can see that their form and structure has a special relationship with the background on which it sits. Not only does the shape of the letter itself give it form, but the space around it*

Individual letters make words and words make up sentences (or lines) and lines make up paragraphs. Paragraphs make up columns of text and columns of text make up pages. When handling type we are dealing with designed letterforms. Only very few experienced typographers take the time and considerable effort to design new typefaces. Though over the years thousands of typefaces have been designed, so the choice for us is enormous. When we look at individual typefaces we can see that they each have their own particular characteristics and "personality." We may examine each of *the 26 letters of the alphabet and can see that their form and structure has a special relationship with the background on which it sits. Not only does the shape of the letter itself give*

MEDIUM
AaBbCcDdEeFfGgHhIiJjKkLlMmNnOoPpQqRrSsTt
UuVvWwXxYyZz1234567890;:'"!?@#£$&*()+=

ITALIC
AaBbCcDdEeFfGgHhIiJjKkLlMmNnOoPpQqRrSsTt
UuVvWwXxYyZz1234567890;:'"!?@#£$&()+=*

DEMIBOLD
AaBbCcDdEeFfGgHhIiJjKkLlMmNnOoPpQqRrSsTt
UuVvWwXxYyZz1234567890;:'"!?@#£$&*()+=

DEMIBOLD ITALIC
AaBbCcDdEeFfGgHhIiJjKkLlMmNnOoPpQqRrSsTt
UuVvWwXxYyZz1234567890;:'"!?@#£$&*()+=

BOLD
AaBbCcDdEeFfGgHhIiJjKkLlMmNnOoPpQqRrSsTt
UuVvWwXxYyZz1234567890;:'"!?@#£$&*()+=

BOLD ITALIC
AaBbCcDdEeFfGgHhIiJjKkLlMmNnOoPpQqRrSsTt
UuVvWwXxYyZz1234567890;:'"!?@#£$&*()+=

ULTRA BLACK
AaBbCcDdEeFfGgHhIiJjKkLlMmNnOoPpQqRrSsTt
UuVvWwXxYyZz1234567890;:'"!?@#£$&*()+=

ULTRA BLACK ITALIC
AaBbCcDdEeFfGgHhIiJjKkLlMmNnOoPpQqRrSsTt
UuVvWwXxYyZz1234567890;:'"!?@#£$&*()+=

SANS SERIF

Syntax

Designed by Hans Edouard Meier, a Swiss type designer and calligrapher, Syntax was first issued in three weights by the Stempel foundry in 1969. Syntax is an influential Humanist sans serif that is a subtle mix of historical form and modernity. It is distinctive in that it has an optically balanced appearance with minimal weight change in the strokes, yet retains the dynamics of the written form. The arches in the lowercase letters show the asymmetric dynamic reminiscent of the pen form, unlike those of a more geometric sans serif. Unusually, all the stroke ends finish at right angles to the direction of the strokes on both the capital and lowercase alphabets. This feature gives it an open and rhythmic liveliness across a line of type. The capitals are slightly shorter than the tall ascenders, which helps to make Syntax highly legible and easy to read. The italics are influenced by the cursive form.

Syntax is a distinctive and extremely versatile typeface that lends itself to a wide range of contexts, ranging from display to complex technical information, on paper and on screen.

Syntax was digitized by Adobe in 1989 and that version is available in four weights of roman and one italic. More recently, with Meier's complete involvement, Syntax has been reworked to include an extended family for wider applications and issued as Linotype Syntax. This is one of the first complete reworkings to have the @ character redesigned to harmonize with the lowercase, making the typographic detailing of e-mail addresses unnecessary. The new feature of a minimal incline to all the roman weights has also been incorporated. Linotype Syntax is available in Regular, Light, Bold, Medium, Heavy, and Black. New small caps for the three lightest weights are available together with an unusual feature —x-height aligning numerals—useful for complex tabular work.

TYPE CHARACTERISTICS

In the capitals, the right angled stroke endings are noticeable in the "N," "V," "W," and "X." In the lowercase letters, the "a" and "g" are two-story; the "s" and "t" are quite wide, the "t" is also tall.

ABCDEFGHIJKLMNOPQRSTUVWXYZ
abcdefghijklmnopqrstuvwxyz1234567890;:'"!?@#£$&*()+=

Quark Default H&J

		Min	Opt	Max
W	⏚	85	110	250
C	⏚	0	0	4
H	✓	ON		
T		0		

Individual letters make words and words make up sentences (or lines) and lines make up paragraphs. Paragraphs make up columns of text and columns of text make up pages. When handling type we are dealing with designed letterforms. Only very few experienced typographers take the time and considerable effort to design new typefaces. Though over the years thousands of typefaces have been designed, so the choice for us is enormous. When we look at individual typefaces we can see that they each have their *own particular characteristics and "personality." We may examine each of the 26 letters of the alphabet and can see that their form and structure has a special relation-*

User Defined H&J

		Min	Opt	Max
W	⏚	60	100	120
C	⏚	–3	0	5
H	✓	ON		
T	✓	3		

Individual letters make words and words make up sentences (or lines) and lines make up paragraphs. Paragraphs make up columns of text and columns of text make up pages. When handling type we are dealing with designed letterforms. Only very few experienced typographers take the time and considerable effort to design new typefaces. Though over the years thousands of typefaces have been designed, so the choice for us is enormous. When we look at individual typefaces we can see that they each have their *own particular characteristics and "personality." We may examine each of the 26 letters of the alphabet and can see that their form and structure has a special relation-*

User Defined H&J

		Min	Opt	Max
W	⏚	70	100	125
C	⏚	0	0	2
H	✓	ON		
T		0		

Individual letters make words and words make up sentences (or lines) and lines make up paragraphs. Paragraphs make up columns of text and columns of text make up pages. When handling type we are dealing with designed letterforms. Only very few experienced typographers take the time and considerable effort to design new typefaces. Though over the years thousands of typefaces have been designed, so the choice for us is enormous. When we look at individual typefaces we can see that they each have their own particular character-*istics and "personality." We may examine each of the 26 letters of the alphabet and can see that their form and structure has a special relationship with the background*

Quark Default H&J

	Min	Opt	Max
W	85	110	250
C	0	0	4
H	ON		
T	0		

Individual letters make words and words make up sentences (or lines) and lines make up paragraphs. Paragraphs make up columns of text and columns of text make up pages. When handling type we are dealing with designed letterforms. Only very few experienced *typographers take the time and considerable effort to design new typefaces. Though*

Quark Default H&J

	Min	Opt	Max
W	85	110	250
C	0	0	4
H	OFF		
T	0		

Individual letters make words and words make up sentences (or lines) and lines make up paragraphs. Paragraphs make up columns of text and columns of text make up pages. When handling type we are dealing with designed letterforms. Only very few experienced *typographers take the time and considerable effort to design new typefaces. Though*

Quark Default H&J

	Min	Opt	Max
W	85	110	250
C	0	0	4
H	ON		
T	0		

Individual letters make words and words make up sentences (or lines) and lines make up paragraphs. Paragraphs make up columns of text and columns of text make up pages. When handling type we are dealing with designed letterforms. Only *very few experienced typographers take the time and considerable effort to design*

Quark Default H&J

	Min	Opt	Max
W	85	110	250
C	0	0	4
H	ON	HZ 14mm	
T	0		

Individual letters make words and words make up sentences (or lines) and lines make up paragraphs. Paragraphs make up columns of text and columns of text make up pages. When handling type we are dealing with designed letterforms. Only *very few experienced typographers take the time and considerable effort to*

User Defined H&J

	Min	Opt	Max
W	70	90	115
C	-3	0	5
H	ON		
T	0		

Individual letters make words and words make up sentences (or lines) and lines make up paragraphs. Paragraphs make up columns of text and columns of text make up pages. When handling type we are dealing with designed letterforms. Only very few experienced typographers *take the time and considerable effort to design new typefaces. Though over the years thou-*

User Defined H&J

	Min	Opt	Max
W	60	90	90
C	-5	0	10
H	ON		
T	2		

Individual letters make words and words make up sentences (or lines) and lines make up paragraphs. Paragraphs make up columns of text and columns of text make up pages. When handling type we are dealing with designed letterforms. Only very few experienced typographers *take the time and considerable effort to design new typefaces. Though over the years thou-*

User Defined H&J

	Min	Opt	Max
W	85	100	250
C	0	0	0
H	OFF		
T	0		

Individual letters make words and words make up sentences (or lines) and lines make up paragraphs. Paragraphs make up columns of text and columns of text make up pages. When handling type we are dealing with designed letterforms. Only very few experienced typographers take the time and considerable effort to design new typefaces. Though over the years thousands of typefaces have been designed, so the choice for us is enormous. When we look at individual typefaces we can see that they each have their own particular characteristics and "personality." We may examine each of the *26 letters of the alphabet and can see that their form and structure has a special relationship with the background on which it sits. Not only does the shape of the letter itself give it*

User Defined H&J

	Min	Opt	Max
W	70	90	110
C	-6	-3	0
H	OFF		
T	0		

Individual letters make words and words make up sentences (or lines) and lines make up paragraphs. Paragraphs make up columns of text and columns of text make up pages. When handling type we are dealing with designed letterforms. Only very few experienced typographers take the time and considerable effort to design new typefaces. Though over the years thousands of typefaces have been designed, so the choice for us is enormous. When we look at individual typefaces we can see that they each have their own particular characteristics and "personality." We may *examine each of the 26 letters of the alphabet and can see that their form and structure has a special relationship with the background on which it sits. Not only does the shape of*

LIGHT AaBbCcDdEeFfGgHhIiJjKkLlMmNnOoPpQqRrSsTt
UuVvWwXxYyZz1234567890;:'"!?@#£$*()+=

LIGHT ITALIC *AaBbCcDdEeFfGgHhIiJjKkLlMmNnOoPpQqRrSsTt
UuVvWwXxYyZz1234567890;:'"!?@#£$*()+=*

REGULAR AaBbCcDdEeFfGgHhIiJjKkLlMmNnOoPpQqRrSsTt
UuVvWwXxYyZz1234567890;:'"!?@#£$*()+=

REGULAR ITALIC *AaBbCcDdEeFfGgHhIiJjKkLlMmNnOoPpQqRrSsTt
UuVvWwXxYyZz1234567890;:'"!?@#£$*()+=*

MEDIUM **AaBbCcDdEeFfGgHhIiJjKkLlMmNnOoPpQqRrSsTt
UuVvWwXxYyZz1234567890;:'"!?@#£$*()+=**

MEDIUM ITALIC ***AaBbCcDdEeFfGgHhIiJjKkLlMmNnOoPpQqRrSsTt
UuVvWwXxYyZz1234567890;:'"!?@#£$*()+=***

Syntax Bold

Syntax

Setting Tips

The design is such that manual
kerning is not normally necessary.
Syntax does not benefit greatly from
close spacing or tracking, as this
tends to conceal the intrinsic form
and so makes it harder to read.

The Syntax family has a
rugged character and is suited
to situations where both text
and display setting is required.

Resilient
Durable
Tough
Robust

AaBbCc

ABCDEFGHIJKLMNOPQRSTUVWXYZ
abcdefghijklmnopqrstuvwxyz1234567890;:'"!?@#£$&*()+=

Quark Default H&J

	Min	Opt	Max
W ✐	85	110	250
C ✐	0	0	4
H ✓	ON		
T	0		

**Individual letters make words and words make up
sentences (or lines) and lines make up paragraphs.
Paragraphs make up columns of text and columns of
text make up pages. When handling type we are
dealing with designed letterforms. Only very few
experienced typographers take the time and considerable effort to design new typefaces. Though over
the years thousands of typefaces have been
designed, so the choice for us is enormous. When we
look at individual typefaces we can see that they
*each have their own particular characteristics and "per-
sonality." We may examine each of the 26 letters of
the alphabet and can see that their form and structure***

User Defined H&J

	Min	Opt	Max
W ✐	80	105	150
C ✐	0	3	3
H ✓	ON		
T ✓	3		

**Individual letters make words and words make up
sentences (or lines) and lines make up paragraphs. Paragraphs make up columns of text and
columns of text make up pages. When handling
type we are dealing with designed letterforms.
Only very few experienced typographers take the
time and considerable effort to design new typefaces. Though over the years thousands of typefaces have been designed, so the choice for us is
enormous. When we look at individual typefaces
*we can see that they each have their own particular
characteristics and "personality." We may examine
each of the 26 letters of the alphabet and can see***

User Defined H&J

	Min	Opt	Max
W ✐	70	100	120
C ✐	−3	0	5
H ✓	ON		
T	0		

**Individual letters make words and words make up sen-
tences (or lines) and lines make up paragraphs.
Paragraphs make up columns of text and columns of
text make up pages. When handling type we are dealing with designed letterforms. Only very few experienced typographers take the time and considerable
effort to design new typefaces. Though over the years
thousands of typefaces have been designed, so the
choice for us is enormous. When we look at individual typefaces we can see that they each have their own
*particular characteristics and "personality." We may
examine each of the 26 letters of the alphabet and can
see that their form and structure has a special relation-***

	Min	Opt	Max
W	85	110	250
C	0	0	4
H	✓ ON		
T	0		

Individual letters make words and words make up sentences (or lines) and lines make up paragraphs. Paragraphs make up columns of text and columns of text make up pages. When handling type we are dealing with designed letterforms. Only *very few experienced typographers take the time and considerable effort to design*

	Min	Opt	Max
W	85	110	250
C	0	0	4
H	OFF		
T	0		

Individual letters make words and words make up sentences (or lines) and lines make up paragraphs. Paragraphs make up columns of text and columns of text make up pages. When handling type we are dealing with designed letterforms. Only *very few experienced typographers take the time and considerable effort to*

	Min	Opt	Max
W	85	110	250
C	0	0	4
H	✓ ON		
T	0		

Individual letters make words and words make up sentences (or lines) and lines make up paragraphs. Paragraphs make up columns of text and columns of text make up pages. When handling type we are dealing with designed letterforms. Only *very few experienced typographers take the time and considerable effort to*

	Min	Opt	Max
W	85	110	250
C	0	0	4
H	✓ ON HZ 14mm		
T	0		

Individual letters make words and words make up sentences (or lines) and lines make up paragraphs. Paragraphs make up columns of text and columns of text make up pages. When handling type we are dealing with designed letterforms. Only *very few experienced typographers take the time and considerable effort to*

	Min	Opt	Max
W	60	90	90
C	−5	0	10
H	✓ ON		
T	0		

Individual letters make words and words make up sentences (or lines) and lines make up paragraphs. Paragraphs make up columns of text and columns of text make up pages. When handling type we are dealing with designed letterforms. Only very few experienced *typographers take the time and considerable effort to design new typefaces. Though*

	Min	Opt	Max
W	70	110	250
C	−3	5	10
H	✓ ON		
T	✓ 2		

Individual letters make words and words make up sentences (or lines) and lines make up paragraphs. Paragraphs make up columns of text and columns of text make up pages. When handling type we are dealing with designed letterforms. Only *very few experienced typographers take the time and considerable effort to*

	Min	Opt	Max
W	85	100	250
C	0	0	0
H	OFF		
T	0		

Individual letters make words and words make up sentences (or lines) and lines make up paragraphs. Paragraphs make up columns of text and columns of text make up pages. When handling type we are dealing with designed letterforms. Only very few experienced typographers take the time and considerable effort to design new typefaces. Though over the years thousands of typefaces have been designed, so the choice for us is enormous. When we look at individual typefaces we can see that they each have their own particular characteristics *and "personality." We may examine each of the 26 letters of the alphabet and can see that their form and structure has a special relationship with the background on which it site. Not*

	Min	Opt	Max
W	60	100	120
C	−3	0	5
H	OFF		
T	0		

Individual letters make words and words make up sentences (or lines) and lines make up paragraphs. Paragraphs make up columns of text and columns of text make up pages. When handling type we are dealing with designed letterforms. Only very few experienced typographers take the time and considerable effort to design new typefaces. Though over the years thousands of typefaces have been designed, so the choice for us is enormous. When we look at individual typefaces we can see that they *each have their own particular characteristics and "personality." We may examine each of the 26 letters of the alphabet and can see that their form and structure*

BOLD AaBbCcDdEeFfGgHhIiJjKkLlMmNnOoPpQqRrSsTt
UuVvWwXxYyZz1234567890;:'"!?@#£$*()+=

BOLD ITALIC *AaBbCcDdEeFfGgHhIiJjKkLlMmNnOoPpQqRrSsTt
UuVvWwXxYyZz1234567890;:'"!?@#£$*()+=*

HEAVY AaBbCcDdEeFfGgHhIiJjKkLlMmNnOoPpQqRrSsTt
UuVvWwXxYyZz1234567890;:'"!?@#£$*()+=

HEAVY ITALIC *AaBbCcDdEeFfGgHhIiJjKkLlMmNnOoPpQqRrSsTt
UuVvWwXxYyZz1234567890;:'"!?@#£$*()+=*

BLACK AaBbCcDdEeFfGgHhIiJjKkLlMmNnOoPpQqRrSsTt
UuVvWwXxYyZz1234567890;:'"!?@#£$*()+=

BLACK ITALIC *AaBbCcDdEeFfGgHhIiJjKkLlMmNnOoPpQqRrSsTt
UuVvWwXxYyZz1234567890;:'"!?@#£$*()+=*

Univers

Univers

Univers, one of the first designs by the leading Swiss type designer Adrian Frutiger, is probably one of the most widely licensed typefaces. Issued by the French type foundry Deberny & Peignot in the late 1950s at the same time as its "rival," Helvetica, Univers has become one of the most significant designs of our time. The design of Univers is modular, which ensures visual compatibility within the unusually extensive family of 21 variations. Frutiger devised a numerical system to identify each variation in which the first number refers to the particular weight and the second to the proportional style of letter: Regular, Condensed, Extended, etc. Roman or italic style is indicated by odd or even second numbers respectively. This system also has built-in quality control, doing away with the confusing use of different names for the same weight by different foundries.

Univers has a greater stroke variation than Helvetica, creating color on the page, and it is comfortable and straightforward to read in lengthy text settings.

TYPE CHARACTERISTICS

The character of Univers is in the detail: the subtle contrast in the stroke weight and the minimally squared curves of the round letters. The design has short ascenders and descenders and is more condensed than Helvetica, but the ends of the curved strokes finish horizontally as in Helvetica. Its visual uniformity gives the design its own individual character, investing it with a subtle interest, high legibility, and readability.

Among the capital letter characteristics are the spur-less "G" with a relatively long stem, and a "Q" entirely different from Helvetica with a flat tail that sits horizontally on the baseline, while the capital "K" has a single junction, as does lowercase "k."

In the lowercase, the two-story "a" has a distinguishing trait in the minimal curve of the straight-backed stem. The "y" has a straight tail, differing from the curved Helvetica tail.

ABCDEFGHIJKLMNOPQRSTUVWXYZ
abcdefghijklmnopqrstuvwxyz1234567890;:'"!?@#£$&*()+=

Quark Default H&J

	Min	Opt	Max
W	85	110	250
C	0	0	4
H	ON		
T	0		

Individual letters make words and words make up sentences (or lines) and lines make up paragraphs. Paragraphs make up columns of text and columns of text make up pages. When handling type we are dealing with designed letterforms. Only very few experienced typographers take the time and considerable effort to design new typefaces. Though over the years thousands of typefaces have been designed, so the choice for us is enormous. When we look at individual typefaces we can see that they *each have their own particular characteristics and "personality." We may examine each of the 26 letters of the alphabet and can see that their form and*

User Defined H&J

	Min	Opt	Max
W	90	100	100
C	0	0	3
H	ON		
T	3		

Individual letters make words and words make up sentences (or lines) and lines make up paragraphs. Paragraphs make up columns of text and columns of text make up pages. When handling type we are dealing with designed letterforms. Only very few experienced typographers take the time and considerable effort to design new typefaces. Though over the years thousands of typefaces have been designed, so the choice for us is enormous. When we look at individual typefaces we can see that they *each have their own particular characteristics and "personality." We may examine each of the 26 letters of the alphabet and can see that their form and*

User Defined H&J

	Min	Opt	Max
W	70	100	105
C	0	0	3
H	ON		
T	0		

Individual letters make words and words make up sentences (or lines) and lines make up paragraphs. Paragraphs make up columns of text and columns of text make up pages. When handling type we are dealing with designed letterforms. Only very few experienced typographers take the time and considerable effort to design new typefaces. Though over the years thousands of typefaces have been designed, so the choice for us is enormous. When we look at individual typefaces we can see that they each have *their own particular characteristics and "personality." We may examine each of the 26 letters of the alphabet and can see that their form and structure*

Quark Default H&J

	Min	Opt	Max
W	85	110	250
C	0	0	4
H	✓ ON		
T	0		

Individual letters make words and words make up sentences (or lines) and lines make up paragraphs. Paragraphs make up columns of text and columns of text make up pages. When handling type we are dealing with designed letterforms. Only *very few experienced typographers take the time and considerable effort to*

Quark Default H&J

	Min	Opt	Max
W	85	110	250
C	0	0	4
H	OFF		
T	0		

Individual letters make words and words make up sentences (or lines) and lines make up paragraphs. Paragraphs make up columns of text and columns of text make up pages. When handling type we are dealing with designed letterforms. Only *very few experienced typographers take the time and considerable effort to*

Quark Default H&J

	Min	Opt	Max
W	85	110	250
C	0	0	4
H	✓ ON		
T	0		

Individual letters make words and words make up sentences (or lines) and lines make up paragraphs. Paragraphs make up columns of text and columns of text make up pages. When handling type we are dealing with designed letterforms. Only *very few experienced typographers take the time and considerable effort to*

Quark Default H&J

	Min	Opt	Max
W	85	110	250
C	0	0	4
H	✓ ON HZ 14mm		
T	0		

Individual letters make words and words make up sentences (or lines) and lines make up paragraphs. Paragraphs make up columns of text and columns of text make up pages. When handling type we are dealing with designed letterforms. Only *very few experienced typographers take the time and considerable effort to*

User Defined H&J

	Min	Opt	Max
W	70	90	115
C	-3	0	5
H	✓ ON		
T	0		

Individual letters make words and words make up sentences (or lines) and lines make up paragraphs. Paragraphs make up columns of text and columns of text make up pages. When handling type we are dealing with designed letterforms. Only very few experienced *typographers take the time and considerable effort to design new typefaces.*

User Defined H&J

	Min	Opt	Max
W	60	90	90
C	-5	0	10
H	✓ ON		
T	✓ 2		

Individual letters make words and words make up sentences (or lines) and lines make up paragraphs. Paragraphs make up columns of text and columns of text make up pages. When handling type we are dealing with designed letterforms. Only very few experienced *typographers take the time and considerable effort to design new typefaces.*

User Defined H&J

	Min	Opt	Max
W	85	100	250
C	0	0	0
H	OFF		
T	0		

Individual letters make words and words make up sentences (or lines) and lines make up paragraphs. Paragraphs make up columns of text and columns of text make up pages. When handling type we are dealing with designed letterforms. Only very few experienced typographers take the time and considerable effort to design new typefaces. Though over the years thousands of typefaces have been designed, so the choice for us is enormous. When we look at individual typefaces we can see that they each have their own particular characteristics *and "personality." We may examine each of the 26 letters of the alphabet and can see that their form and structure has a special relationship with the background on which it*

User Defined H&J

	Min	Opt	Max
W	70	90	110
C	-6	-3	0
H	OFF		
T	0		

Individual letters make words and words make up sentences (or lines) and lines make up paragraphs. Paragraphs make up columns of text and columns of text make up pages. When handling type we are dealing with designed letterforms. Only very few experienced typographers take the time and considerable effort to design new typefaces. Though over the years thousands of typefaces have been designed, so the choice for us is enormous. When we look at individual typefaces we can see that they each have their own particular *characteristics and "personality." We may examine each of the 26 letters of the alphabet and can see that their form and structure has a special relationship with the*

56 ROMAN ITALIC *AaBbCcDdEeFfGgHhIiJjKkLlMmNnOoPpQqRrSsTt UuVvWwXxYyZz1234567890;:'"!?@#£$&*()+=*

45 LIGHT AaBbCcDdEeFfGgHhIiJjKkLlMmNnOoPpQqRrSsTt UuVvWwXxYyZz1234567890;:'"!?@#£$&*()+=

46 LIGHT ITALIC *AaBbCcDdEeFfGgHhIiJjKkLlMmNnOoPpQqRrSsTt UuVvWwXxYyZz1234567890;:'"!?@#£$&*()+=*

65 BOLD **AaBbCcDdEeFfGgHhIiJjKkLlMmNnOoPpQqRrSsTt UuVvWwXxYyZz1234567890;:'"!?@#£$&*()+=**

66 BOLD ITALIC ***AaBbCcDdEeFfGgHhIiJjKkLlMmNnOoPpQqRrSsTt UuVvWwXxYyZz1234567890;:'"!?@#£$&*()+=***

SANS SERIF

Univers Bold

Univers

Setting Tips

As with some other sans serif faces, Univers benefits from generous leading. The extensive Univers family is highly versatile and lends itself to a wide range of situations, including display, complex hierarchical information that needs a clear, well "signposted" interpretation, and document and report design.

Letters

Further reading: Chapter 16 pages 630–633

Individual letters make words and words make up sentences (or lines) and lines make up paragraphs. Paragraphs make up columns of text and columns of text make up pages. When handling type we are dealing with designed letterforms. Only very few experienced typographers take the time and considerable effort to design new typefaces.

Choice

Further reading: Chapter 16 pages 630–633

Though over the years thousands of typefaces have been designed, so the choice for us is enormous. When we look at individual typefaces we can see that they each have their own particular characteristics and "personality." We may examine each of the 26 letters of the alphabet and can see that their form and structure has a special relationship with the background on which it sits.

ABCDEFGHIJKLMNOPQRSTUVWXYZ
abcdefghijklmnopqrstuvwxyz1234567890;:'"!?@#£$&*()+=

Quark Default H&J

		Min	Opt	Max
W		85	110	250
C		0	0	4
H	✓	ON		
T		0		

Individual letters make words and words make up sentences (or lines) and lines make up paragraphs. Paragraphs make up columns of text and columns of text make up pages. When handling type we are dealing with designed letterforms. Only very few experienced typographers take the time and considerable effort to design new typefaces. Though over the years thousands of typefaces have been designed, so the choice for us is enormous. When we look at individual typefaces we can see that they *each have their own particular characteristics and "personality." We may examine each of the 26 letters of the alphabet and can see that their form and*

User Defined H&J

		Min	Opt	Max
W		75	100	105
C		0	0	0
H	✓	ON		
T	✓	3		

Individual letters make words and words make up sentences (or lines) and lines make up paragraphs. Paragraphs make up columns of text and columns of text make up pages. When handling type we are dealing with designed letterforms. Only very few experienced typographers take the time and considerable effort to design new typefaces. Though over the years thousands of typefaces have been designed, so the choice for us is enormous. When we look at individual typefaces we can see that *they each have their own particular characteristics and "personality." We may examine each of the 26 letters of the alphabet and can see that their form*

User Defined H&J

		Min	Opt	Max
W		70	100	120
C		–1	0	3
H	✓	ON		
T		0		

User Defined H&J

		Min	Opt	Max
W		70	90	110
C		–6	–3	0
H		OFF		
T		0		

Individual letters make words and words make up sentences (or lines) and lines make up paragraphs. Paragraphs make up columns of text and columns of text make up pages. When handling type we are dealing with designed letterforms. Only very few experienced typographers take the time and considerable

Individual letters make words and words make up sentences (or lines) and lines make up paragraphs. Paragraphs make up columns of text and columns of text make up pages. When handling type we are dealing with designed letterforms. Only very few experienced typographers take the time and considerable effort to

Quark Default H&J

	Min	Opt	Max
W ✐	85	110	250
C ✐	0	0	4
H	OFF		
T	0		

Individual letters make words and words make up sentences (or lines) and lines make up paragraphs. Paragraphs make up columns of text and columns of text make up pages. When handling type we are dealing with designed letterforms. Only very few experienced typographers take the time and considerable effort to design new typefaces. Though over the years thousands of typefaces have been designed, so the choice for us is enormous. When we look at individual typefaces *we can see that they each have their own particular characteristics and "personality." We may examine each of the 26 letters of the alphabet*

User Defined H&J

	Min	Opt	Max
W ✐	85	100	250
C ✐	0	0	0
H	OFF		
T	0		

Individual letters make words and words make up sentences (or lines) and lines make up paragraphs. Paragraphs make up columns of text and columns of text make up pages. When handling type we are dealing with designed letterforms. Only very few experienced typographers take the time and considerable effort to design new typefaces. Though over the years thousands of typefaces have been designed, so the choice for us is enormous. When we look at individual typefaces *we can see that they each have their own particular characteristics and "personality." We may examine each of the 26 letters of the alphabet*

Univers

⌐14   ⌐15

Univers Roman and Bold showing overall tracking of −12 with letter pairs individually kerned as shown. Gray text indicates 0 tracking and no kerning of letter pairs.

Univers

⌐14   ⌐9   ⌐16   ⌐7

75 BLACK
AaBbCcDdEeFfGgHhIiJjKkLlMmNnOoPpQqRrSsTt UuVvWwXxYyZz1234567890;:'"!?@#£$&*()+=

76 BLACK ITALIC
AaBbCcDdEeFfGgHhIiJjKkLlMmNnOoPpQqRrSsTt UuVvWwXxYyZz1234567890;:'"!?@#£$&()+=*

85 EXTRA BLACK
AaBbCcDdEeFfGgHhIiJjKkLlMmNnOoPpQqRrSsTt UuVvWwXxYyZz1234567890;:'"!?@#£$&*()+=

86 EXTRA BLACK ITALIC
AaBbCcDdEeFfGgHhIiJjKkLlMmNnOoPpQqRrSsTt UuVvWwXxYyZz1234567890;:'"!?@#£$&()+=*

47 LIGHT COND
AaBbCcDdEeFfGgHhIiJjKkLlMmNnOoPpQqRrSsTt UuVvWwXxYyZz1234567890;:'"!?@#£$&*()+=

48 LIGHT COND ITALIC
AaBbCcDdEeFfGgHhIiJjKkLlMmNnOoPpQqRrSsTt UuVvWwXxYyZz1234567890;:'"!?@#£$&()+=*

57 COND
AaBbCcDdEeFfGgHhIiJjKkLlMmNnOoPpQqRrSsTt UuVvWwXxYyZz1234567890;:'"!?@#£$&*()+=

58 COND ITALIC
AaBbCcDdEeFfGgHhIiJjKkLlMmNnOoPpQqRrSsTt UuVvWwXxYyZz1234567890;:'"!?@#£$&()+=*

67 BOLD COND
AaBbCcDdEeFfGgHhIiJjKkLlMmNnOoPpQqRrSsTt UuVvWwXxYyZz1234567890;:'"!?@#£$&*()+=

68 BOLD COND ITALIC
AaBbCcDdEeFfGgHhIiJjKkLlMmNnOoPpQqRrSsTt UuVvWwXxYyZz1234567890;:'"!?@#£$&()+=*

49 LIGHT ULTRA COND
AaBbCcDdEeFfGgHhIiJjKkLlMmNnOoPpQqRrSsTt UuVvWwXxYyZz1234567890;:'"!?@#£$&*()+=

59 ULTRA COND
AaBbCcDdEeFfGgHhIiJjKkLlMmNnOoPpQqRrSsTt UuVvWwXxYyZz1234567890;:'"!?@#£$&*()+=

69 BOLD ULTRA COND
AaBbCcDdEeFfGgHhIiJjKkLlMmNnOoPpQqRrSsTt UuVvWwXxYyZz1234567890;:'"!?@#£$&*()+=

53 EXT
AaBbCcDdEeFfGgHhIiJjKkLlMmN OoPpQqRrSsTtUuVvWwXxYyZz 1234567890;:'"!?@#£$&*()+=

54 EXT ITALIC
AaBbCcDdEeFfGgHhIiJjKkLlMmN OoPpQqRrSsTtUuVvWwXxYyZz 1234567890;:'"!?@#£$&()+=*

63 BOLD EXT
AaBbCcDdEeFfGgHhIiJjKkLlMmN OoPpQqRrSsTtUuVvWwXxYyZz 1234567890;:'"!?@#£$&*()+=

64 BOLD EXT ITALIC
AaBbCcDdEeFfGgHhIiJjKkLlMmN OoPpQqRrSsTtUuVvWwXxYyZz 1234567890;:'"!?@#£$&()+=*

73 BLACK EXT
AaBbCcDdEeFfGgHhIiJjKkLlMmN OoPpQqRrSsTtUuVvWwXxYyZz 1234567890;:'"!?@#£$&*()+=

74 BLACK EXT ITALIC
AaBbCcDdEeFfGgHhIiJjKkLlMmN OoPpQqRrSsTtUuVvWwXxYyZz 1234567890;:'"!?@#£$&()+=*

83 EXTRA BLACK EXT
AaBbCcDdEeFfGgHhIiJjKkLlMmN OoPpQqRrSsTtUuVvWwXxYyZz 1234567890;:'"!?@#£$&*()+=

84 EXTRA BLACK EXT ITALIC
AaBbCcDdEeFfGgHhIiJjKkLlMmN OoPpQqRrSsTtUuVvWwXxYyZz 1234567890;:'"!?@#£$&()+=*

SANS SERIF

ITC Officina Sans and Serif

Originally conceived by the leading German type designer Erik Spiekermann, as a workhorse for the office, ITC Officina was purposefully designed in two weights, book and bold, in a related serif and sans serif design with italic versions. However, its down-to-earth functional appeal spread far beyond the office. It was "discovered" and used in publicity and editorial contexts as a fashionable understatement. Its new uses demanded an extended family of weights. In response to this unexpected development. Spiekermann, the founder of MetaDesign, worked with the company's typography and type director, Ole Schäfer, at MetaDesign to produce the current family of Medium, Extra Bold, and Black for both the serif and sans styles and their italic versions. Small caps and Old Style (nonlining) figures are available for all weights.

TYPE CHARACTERISTICS / SANS

Other than lacking serifs, the individual letterforms of the sans version have the same weight and proportions of the serif version. The capital "J" retains its half-crossbar and, curiously, the numeral "1" retains all its serifs. Stroke endings are terminated at right angles to the stroke direction, which is particularly noticeable in both the capital and lowercase "k," "v," "w," "x," and "y."

TYPE CHARACTERISTICS / SERIF

Officina Serif has a narrow economical slab serif design. The lowercase "g" is single story throughout the family. In the italics the lowercase "f" has a full-length descender. Capital "M" has a high center junction. The capital "J" has an unusual half-crossbar to the top, as do the lowercase "i" and "j" in both the roman and the italic versions. The numerals are identical to the sans serif version, i.e., no serifs except on numeral "1." Apart from a single-story lowercase "a," a script-like lowercase "f" and "b," the italic is very much like a sloped roman. Lowercase "y" is distinctive owing to the staggered junction between strokes (similar to Meta). Punctuation characters are serif-less.

ABCDEFGHIJKLMNOPQRSTUVWXYZ
abcdefghijklmnopqrstuvwxyz1234567890;:'"!?@#£$&*()+=

Quark Default H&J	Min	Opt	Max
W	85	110	250
C	0	0	4
H	ON		
T	0		

Individual letters make words and words make up sentences (or lines) and lines make up paragraphs. Paragraphs make up columns of text and columns of text make up pages. When handling type we are dealing with designed letterforms. Only very few experienced typographers take the time and considerable effort to design new typefaces. Though over the years thousands of typefaces have been designed, so the choice for us is enormous. When we look at individual typefaces we can see that they each have their own particular characteristics and "personality." We may examine each of the 26 letters of *the alphabet and can see that their form and structure has a special relationship with the background on which it sits. Not only does the shape of the letter itself give it form, but the*

User Defined H&J	Min	Opt	Max
W	70	90	90
C	3	3	4
H	ON		
T	3		

Individual letters make words and words make up sentences (or lines) and lines make up paragraphs. Paragraphs make up columns of text and columns of text make up pages. When handling type we are dealing with designed letterforms. Only very few experienced typographers take the time and considerable effort to design new typefaces. Though over the years thousands of typefaces have been designed, so the choice for us is enormous. When we look at individual typefaces we can see that they each have their own particular characteristics and "personality." We *may examine each of the 26 letters of the alphabet and can see that their form and structure has a special relationship with the background on which it sits. Not only does the*

User Defined H&J	Min	Opt	Max
W	60	100	120
C	-3	0	5
H	ON		
T	0		

Individual letters make words and words make up sentences (or lines) and lines make up paragraphs. Paragraphs make up columns of text and columns of text make up pages. When handling type we are dealing with designed letterforms. Only very few experienced typographers take the time and considerable effort to design new typefaces. Though over the years thousands of typefaces have been designed, so the choice for us is enormous. When we look at individual typefaces we can see that they each have their own particular characteristics and "personality." We may examine each of the 26 letters of the *alphabet and can see that their form and structure has a special relationship with the background on which it sits. Not only does the shape of the letter itself give it form, but the space*

Quark Default H&J

	Min	Opt	Max
W	85	110	250
C	0	0	4
H ✓	ON		
T	0		

Individual letters make words and words make up sentences (or lines) and lines make up paragraphs. Paragraphs make up columns of text and columns of text make up pages. When handling type we are dealing with designed letterforms. Only very few experienced typographers take the time and considerable *effort to design new typefaces. Though over the years thousands of typefaces have been designed,*

Quark Default H&J

	Min	Opt	Max
W	85	110	250
C	0	0	4
H	OFF		
T	0		

Individual letters make words and words make up sentences (or lines) and lines make up paragraphs. Paragraphs make up columns of text and columns of text make up pages. When handling type we are dealing with designed letterforms. Only very few experienced typographers take the time and *considerable effort to design new typefaces. Though over the years thousands of typefaces have*

Quark Default H&J

	Min	Opt	Max
W	85	110	250
C	0	0	4
H ✓	ON		
T	0		

Individual letters make words and words make up sentences (or lines) and lines make up paragraphs. Paragraphs make up columns of text and columns of text make up pages. When handling type we are dealing with designed letterforms. Only very few experienced typographers take *the time and considerable effort to design new typefaces. Though over the years thou-*

Quark Default H&J

	Min	Opt	Max
W	85	110	250
C	0	0	4
H ✓	ON HZ 14mm		
T	0		

Individual letters make words and words make up sentences (or lines) and lines make up paragraphs. Paragraphs make up columns of text and columns of text make up pages. When handling type we are dealing with designed letterforms. Only very few experienced typographers take *the time and considerable effort to design new typefaces. Though over the years*

User Defined H&J

	Min	Opt	Max
W	60	90	90
C	-5	0	10
H ✓	ON		
T	0		

Individual letters make words and words make up sentences (or lines) and lines make up paragraphs. Paragraphs make up columns of text and columns of text make up pages. When handling type we are dealing with designed letterforms. Only very few experienced typographers take the time and considerable effort to design *new typefaces. Though over the years thousands of typefaces have been designed, so the choice for us*

User Defined H&J

	Min	Opt	Max
W	75	90	110
C	-3	-3	3
H ✓	ON		
T ✓	2		

Individual letters make words and words make up sentences (or lines) and lines make up paragraphs. Paragraphs make up columns of text and columns of text make up pages. When handling type we are dealing with designed letterforms. Only very few experienced typographers take the time and considerable *effort to design new typefaces. Though over the years thousands of typefaces have been designed,*

User Defined H&J

	Min	Opt	Max
W	85	100	250
C	0	0	0
H	OFF		
T	0		

Individual letters make words and words make up sentences (or lines) and lines make up paragraphs. Paragraphs make up columns of text and columns of text make up pages. When handling type we are dealing with designed letterforms. Only very few experienced typographers take the time and considerable effort to design new typefaces. Though over the years thousands of typefaces have been designed, so the choice for us is enormous. When we look at individual typefaces we can see that they each have their own particular characteristics and "personality." We may examine each of the 26 letters of the alphabet and can see that their form and *structure has a special relationship with the background on which it sits. Not only does the shape of the letter itself give it form, but the space around it accents its individuality. When two or more letters are*

User Defined H&J

	Min	Opt	Max
W	70	100	120
C	-3	0	5
H	OFF		
T	0		

Individual letters make words and words make up sentences (or lines) and lines make up paragraphs. Paragraphs make up columns of text and columns of text make up pages. When handling type we are dealing with designed letterforms. Only very few experienced typographers take the time and considerable effort to design new typefaces. Though over the years thousands of typefaces have been designed, so the choice for us is enormous. When we look at individual typefaces we can see that they each have their own particular characteristics and "personality." We may examine each of the 26 letters of *the alphabet and can see that their form and structure has a special relationship with the background on which it sits. Not only does the shape of the letter itself give it form, but the space*

SANS ITALIC *AaBbCcDdEeFfGgHhIiJjKkLlMmNnOoPpQqRrSsTt UuVvWwXxYyZz1234567890;:'"!?@#£$&*()+=*

SANS BOLD **AaBbCcDdEeFfGgHhIiJjKkLlMmNnOoPpQqRrSsTt UuVvWwXxYyZz1234567890;:'"!?@#£$&*()+=**

SANS BOLD ITALIC ***AaBbCcDdEeFfGgHhIiJjKkLlMmNnOoPpQqRrSsTt UuVvWwXxYyZz1234567890;:'"!?@#£$&*()+=***

ITC Officina Serif and Sans

Officina

Setting Tips

A relatively narrow face, Officina can be used where economy of space is important and can accept a small amount of minus-tracking.

ABCDEFGHIJKLMN
ABCDEFGHIJKLMN
OPQRSTUVWXYZ
OPQRSTUVWXYZ
abcdefghijklmn
abcdefghijklmn
opqrstuvwxyz
opqrstuvwxyz

Officina Serif and Officina Sans share the same rhythms and weight, yet the Serif adds more ink onto the printed sheet— useful when working with colored type.

AaBbCc
ABCDEFGHIJKLMNOPQRSTUVWXYZ
abcdefghijklmnopqrstuvwxyz1234567890;:'"!?@#£$&*()+=

Quark Default H&J

		Min	Opt	Max
W	✐	85	110	250
C	✐	0	0	4
H	✓	ON		
T		0		

Individual letters make words and words make up sentences (or lines) and lines make up paragraphs. Paragraphs make up columns of text and columns of text make up pages. When handling type we are dealing with designed letterforms. Only very few experienced typographers take the time and considerable effort to design new typefaces. Though over the years thousands of typefaces have been designed, so the choice for us is enormous. When we look at individual typefaces we can see that they each have their own particular characteristics and "personality." We may *examine each of the 26 letters of the alphabet and can see that their form and structure has a special relationship with the background on which it sits. Not only does the shape of*

User Defined H&J

		Min	Opt	Max
W	✐	70	100	105
C	✐	0	0	3
H	✓	ON		
T	✓	3		

Individual letters make words and words make up sentences (or lines) and lines make up paragraphs. Paragraphs make up columns of text and columns of text make up pages. When handling type we are dealing with designed letterforms. Only very few experienced typographers take the time and considerable effort to design new typefaces. Though over the years thousands of typefaces have been designed, so the choice for us is enormous. When we look at individual typefaces we can see that they each have their own particular characteristics and "personality." We *may examine each of the 26 letters of the alphabet and can see that their form and structure has a special relationship with the background on which it sits. Not only does the*

User Defined H&J

		Min	Opt	Max
W	✐	70	90	110
C	✐	−6	−3	0
H	✓	ON		
T		0		

Individual letters make words and words make up sentences (or lines) and lines make up paragraphs. Paragraphs make up columns of text and columns of text make up pages. When handling type we are dealing with designed letterforms. Only very few experienced typographers take the time and considerable effort to design new typefaces. Though over the years thousands of typefaces have been designed, so the choice for us is enormous. When we look at individual typefaces we can see that they each have their own particular characteristics and "personality." We may examine each of the 26 letters of the *alphabet and can see that their form and structure has a special relationship with the background on which it sits. Not only does the shape of the letter itself give it form, but the space around*

Quark Default H&J	Min	Opt	Max
W	85	110	250
C	0	0	4
H ✓	ON		
T	0		

Individual letters make words and words make up sentences (or lines) and lines make up paragraphs. Paragraphs make up columns of text and columns of text make up pages. When handling type we are dealing with designed letterforms. Only very few experienced typographers take the *time and considerable effort to design new typefaces. Though over the years thousands of*

Quark Default H&J	Min	Opt	Max
W	85	110	250
C	0	0	4
H	OFF		
T	0		

Individual letters make words and words make up sentences (or lines) and lines make up paragraphs. Paragraphs make up columns of text and columns of text make up pages. When handling type we are dealing with designed letterforms. Only very few experienced typographers take *the time and considerable effort to design new typefaces. Though over the years*

Quark Default H&J	Min	Opt	Max
W	85	110	250
C	0	0	4
H ✓	ON		
T	0		

Individual letters make words and words make up sentences (or lines) and lines make up paragraphs. Paragraphs make up columns of text and columns of text make up pages. When handling type we are dealing with designed letterforms. Only very few experienced typographers take *the time and considerable effort to design new typefaces. Though over the years thou-*

Quark Default H&J	Min	Opt	Max
W	85	110	250
C	0	0	4
H ✓	ON HZ 14mm		
T	0		

Individual letters make words and words make up sentences (or lines) and lines make up paragraphs. Paragraphs make up columns of text and columns of text make up pages. When handling type we are dealing with designed letterforms. Only very few experienced typographers take *the time and considerable effort to design new typefaces. Though over the years*

User Defined H&J	Min	Opt	Max
W	60	90	110
C	–5	–5	5
H ✓	ON		
T	0		

Individual letters make words and words make up sentences (or lines) and lines make up paragraphs. Paragraphs make up columns of text and columns of text make up pages. When handling type we are dealing with designed letterforms. Only very few experienced typographers take the time and considerable *effort to design new typefaces. Though over the years thousands of typefaces have been designed,*

User Defined H&J	Min	Opt	Max
W	70	90	115
C	–3	0	5
H ✓	ON		
T ✓	2		

Individual letters make words and words make up sentences (or lines) and lines make up paragraphs. Paragraphs make up columns of text and columns of text make up pages. When handling type we are dealing with designed letterforms. Only very few experienced typographers take the time and con-*siderable effort to design new typefaces. Though over the years thousands of typefaces have*

User Defined H&J	Min	Opt	Max
W	85	100	250
C	0	0	0
H	OFF		
T	0		

Individual letters make words and words make up sentences (or lines) and lines make up paragraphs. Paragraphs make up columns of text and columns of text make up pages. When handling type we are dealing with designed letterforms. Only very few experienced typographers take the time and considerable effort to design new typefaces. Though over the years thousands of typefaces have been designed, so the choice for us is enormous. When we look at individual typefaces we can see that they each have their own particular characteristics and "personality." We may examine each of the 26 letters of the alphabet and can see that *their form and structure has a special relationship with the background on which it sits. Not only does the shape of the letter itself give it form, but the space around it accents its individuality.*

User Defined H&J	Min	Opt	Max
W	70	100	120
C	–3	0	5
H	OFF		
T	0		

Individual letters make words and words make up sentences (or lines) and lines make up paragraphs. Paragraphs make up columns of text and columns of text make up pages. When handling type we are dealing with designed letterforms. Only very few experienced typographers take the time and considerable effort to design new typefaces. Though over the years thousands of typefaces have been designed, so the choice for us is enormous. When we look at individual typefaces we can see that they each have their own particular characteristics and "personality." We may examine each of *the 26 letters of the alphabet and can see that their form and structure has a special relationship with the background on which it sits. Not only does the shape of the letter itself give*

SERIF ITALIC *AaBbCcDdEeFfGgHhIiJjKkLlMmNnOoPpQqRrSsTt UuVvWwXxYyZz1234567890;:'"!?@#£$&*()+=*

SERIF BOLD **AaBbCcDdEeFfGgHhIiJjKkLlMmNnOoPpQqRrSsTt UuVvWwXxYyZz1234567890;:'"!?@#£$&*()+=**

SERIF BOLD ITALIC ***AaBbCcDdEeFfGgHhIiJjKkLlMmNnOoPpQqRrSsTt UuVvWwXxYyZz1234567890;:'"!?@#£$&*()+=***

Rotis Sans and Serif

The Rotis family is unique in that it includes secondary designs related to both its sans and serif versions. Otl Aicher, the German graphic and type designer and author, best known for his set of Olympic pictograms for the 1972 Munich Games, designed this unusual type family for Agfa in 1989. Named after the location of his studio, Rotis was to be Aicher's last typeface design.

Rotis is a measured, rhythmic functional design with a good visual flow. The four family designs have common proportions, weights, and heights so can be cohesively combined with each other, either to complement or contrast, making Rotis suited to a wide range of coordinated contexts, including display and continuous text.

TYPE CHARACTERISTICS

Rotis is easily recognizable by its quirky capital and lowercase "c." In all four versions of both the capital and lowercase alphabets, it has a distinct backward direction and open lower curve, as does the lowercase "e." The ampersand ligature probably resembles its Latin origin—the word "*Et*," more than any other typeface.

The Semi Sans version has a stroke weight change (reminiscent of Optima). The Serif version has the same stroke weight change and, apart from the "a," "e," "f," "o," "q," and "s," all the lowercase letters have a single serif, while the "z," has two. Similarly, most capitals have the single serif.

The family is numerically identified and includes Sans Serif, Semi Sans, Serif, and Semi Serif. The Sans Serif and Semi Sans versions are available in 45 Light, 55 Regular, both with italic versions; 65 Bold, and 75 Extra Bold weights without italics. The Serif comes in 55 Regular, with an italic, and 65 Bold without. The Semi Serif version is in 55 Regular and 65 Bold without italics.

ABCDEFGHIJKLMNOPQRSTUVWXYZ
abcdefghijklmnopqrstuvwxyz1234567890;:'"!?@#£$&*()+=

Quark Default H&J

	Min	Opt	Max
W	85	110	250
C	0	0	4
H	ON		
T	0		

Individual letters make words and words make up sentences (or lines) and lines make up paragraphs. Paragraphs make up columns of text and columns of text make up pages. When handling type we are dealing with designed letterforms. Only very few experienced typographers take the time and considerable effort to design new typefaces. Though over the years thousands of typefaces have been designed, so the choice for us is enormous. When we look at individual typefaces we can see that they each have their own particular characteristics and "personality." We may examine each of the 26 letters of the *alphabet and can see that their form and structure has a special relationship with the background on which it sits. Not only does the shape of the letter itself give it form, but the space around it*

User Defined H&J

	Min	Opt	Max
W	90	100	105
C	0	0	3
H	ON		
T	3		

Individual letters make words and words make up sentences (or lines) and lines make up paragraphs. Paragraphs make up columns of text and columns of text make up pages. When handling type we are dealing with designed letterforms. Only very few experienced typographers take the time and considerable effort to design new typefaces. Though over the years thousands of typefaces have been designed, so the choice for us is enormous. When we look at individual typefaces we can see that they each have their own particular characteristics and "personality." We may examine each of the 26 letters of *the alphabet and can see that their form and structure has a special relationship with the background on which it sits. Not only does the shape of the letter itself give it form, but the*

User Defined H&J

	Min	Opt	Max
W	70	85	100
C	-5	-5	0
H	ON		
T	0		

Individual letters make words and words make up sentences (or lines) and lines make up paragraphs. Paragraphs make up columns of text and columns of text make up pages. When handling type we arc dealing with designed letterforms. Only very few experienced typographers take the time and considerable effort to design new typefaces. Though over the years thousands of typefaces have been designed, so the choice for us is enormous. When we look at individual typefaces we can see that they each have their own particular characteristics and "personality." We may examine each of the 26 letters of the alphabet and can see that their form and structure has *a special relationship with the background on which it sits. Not only . does the shape of the letter itself give it form, but the space around it accents its individuality When two or more letters are put side by side*

Quark Default H&J

	Min	Opt	Max
W	85	110	250
C	0	0	4
H	✓ ON		
T	0		

Individual letters make words and words make up sentences (or lines) and lines make up paragraphs. Paragraphs make up columns of text and columns of text make up pages. When handling type we are dealing with designed letterforms. Only very few experienced typographers take the time and considerable *effort to design new typefaces. Though over the years thousands of typefaces have been designed,*

Quark Default H&J

	Min	Opt	Max
W	85	110	250
C	0	0	4
H	OFF		
T	0		

Individual letters make words and words make up sentences (or lines) and lines make up paragraphs. Paragraphs make up columns of text and columns of text make up pages. When handling type we are dealing with designed letterforms. Only very few experienced typographers take the time and *considerable effort to design new typefaces. Though over the years thousands of typefaces have been*

Quark Default H&J

	Min	Opt	Max
W	85	110	250
C	0	0	4
H	✓ ON		
T	0		

Individual letters make words and words make up sentences (or lines) and lines make up paragraphs. Paragraphs make up columns of text and columns of text make up pages. When handling type we are dealing with designed letterforms. Only very few experienced typographers take the time and considerable *effort to design new typefaces. Though over the years thousands of typefaces have been designed,*

Quark Default H&J

	Min	Opt	Max
W	85	110	250
C	0	0	4
H	✓ ON	HZ 14mm	
T	0		

Individual letters make words and words make up sentences (or lines) and lines make up paragraphs. Paragraphs make up columns of text and columns of text make up pages. When handling type we are dealing with designed letterforms. Only very few experienced typographers take the time and *considerable effort to design new typefaces. Though over the years thousands of typefaces have*

User Defined H&J

	Min	Opt	Max
W	60	90	90
C	-5	0	10
H	✓ ON		
T	0		

Individual letters make words and words make up sentences (or lines) and lines make up paragraphs. Paragraphs make up columns of text and columns of text make up pages. When handling type we are dealing with designed letterforms. Only very few experienced typographers take the time and considerable effort to design new typefaces. *Though over the years thousands of typefaces have been designed, so the choice for us is enormous. When*

User Defined H&J

	Min	Opt	Max
W	75	100	115
C	0	0	10
H	✓ ON		
T	✓ 2		

Individual letters make words and words make up sentences (or lines) and lines make up paragraphs. Paragraphs make up columns of text and columns of text make up pages. When handling type we are dealing with designed letterforms. Only very few experienced typographers take the time and considerable *effort to design new typefaces. Though over the years thousands of typefaces have been designed,*

User Defined H&J

	Min	Opt	Max
W	85	100	250
C	0	0	0
H	OFF		
T	0		

Individual letters make words and words make up sentences (or lines) and lines make up paragraphs. Paragraphs make up columns of text and columns of text make up pages. When handling type we are dealing with designed letterforms. Only very few experienced typographers take the time and considerable effort to design new typefaces. Though over the years thousands of typefaces have been designed, so the choice for us is enormous. When we look at individual typefaces we can see that they each have their own particular characteristics and "personality." We may examine each of the 26 letters of the alphabet and can see that their form and structure has a special relationship with the *background on which it sits. Not only does the shape of the letter itself give it form, but the space around it accents its individuality. When two or more letters are put side by side they create words which are*

User Defined H&J

	Min	Opt	Max
W	70	90	110
C	-6	-3	0
H	OFF		
T	0		

Individual letters make words and words make up sentences (or lines) and lines make up paragraphs. Paragraphs make up columns of text and columns of text make up pages. When handling type we are dealing with designed letterforms. Only very few experienced typographers take the time and considerable effort to design new typefaces. Though over the years thousands of typefaces have been designed, so the choice for us is enormous. When we look at individual typefaces we can see that they each have their own particular characteristics and "personality." We may examine each of the 26 letters of the alphabet and can see that their form and *structure has a special relationship with the background on which it sits. Not only does the shape of the letter itself give it form, but the space around it accents its individuality. When two or more letters*

SANS 56 ITALIC	*AaBbCcDdEeFfGgHhIiJjKkLlMmNnOoPpQqRrSsTt UuVvWwXxYyZz1234567890;:'"!?@#£$&*()+=*
SANS 45 LIGHT	AaBbCcDdEeFfGgHhIiJjKkLlMmNnOoPpQqRrSsTt UuVvWwXxYyZz1234567890;:'"!?@#£$&*()+=
SANS 46 LIGHT ITALIC	*AaBbCcDdEeFfGgHhIiJjKkLlMmNnOoPpQqRrSsTt UuVvWwXxYyZz1234567890;:'"!?@#£$&*()+=*
SANS 65 BOLD	**AaBbCcDdEeFfGgHhIiJjKkLlMmNnOoPpQqRrSsTt UuVvWwXxYyZz1234567890;:'"!?@#£$&*()+=**
SANS 75 EXTRA BOLD	**AaBbCcDdEeFfGgHhIiJjKkLlMmNnOoPpQqRrSsTt UuVvWwXxYyZz1234567890;:'"!?@#£$&*()+=**

Rotis Serif and Sans

Rotis

Setting Tips

As with many digital faces, the letterforms work well with their standard tracking values without modification. However, the large x-height of Rotis means an increase in leading in large amounts of text setting is normally necessary.

TRACK –10

Technical

TRACK –7

Legal

TRACK –3

Historical

–7 –10 2

The Rotis family may be used to express different moods. In display sizes, minus-tracking may differ from one style to another.

Despite variations in detailing, it can be seen that the different styles of Rotis share a structural commonality.

AaBbCc

ABCDEFGHIJKLMNOPQRSTUVWXYZ
abcdefghijklmnopqrstuvwxyz1234567890;:'"!?@#£$&*()+=

Quark Default H&J			
	Min	Opt	Max
W	85	110	250
C	0	0	4
H ✓	ON		
T	0		

Individual letters make words and words make up sentences (or lines) and lines make up paragraphs. Paragraphs make up columns of text and columns of text make up pages. When handling type we are dealing with designed letterforms. Only very few experienced typographers take the time and considerable effort to design new typefaces. Though over the years thousands of typefaces have been designed, so the choice for us is enormous. When we look at individual typefaces we can see that they each have their own particular characteristics and "personality." We may *examine each of the 26 letters of the alphabet and can see that their form and structure has a special relationship with the background on which it sits. Not only does the*

User Defined H&J			
	Min	Opt	Max
W	70	100	120
C	–1	0	3
H ✓	ON		
T ✓	3		

Individual letters make words and words make up sentences (or lines) and lines make up paragraphs. Paragraphs make up columns of text and columns of text make up pages. When handling type we are dealing with designed letterforms. Only very few experienced typographers take the time and considerable effort to design new typefaces. Though over the years thousands of typefaces have been designed, so the choice for us is enormous. When we look at individual typefaces we can see that they each have their own particular characteristics and "personality." We *may examine each of the 26 letters of the alphabet and can see that their form and structure has a special relationship with the background on which it sits. Not only does*

User Defined H&J			
	Min	Opt	Max
W	70	85	100
C	–5	–5	0
H ✓	ON		
T	0		

User Defined H&J			
	Min	Opt	Max
W	70	100	120
C	–3	0	5
H	OFF		
T	0		

Individual letters make words and words make up sentences (or lines) and lines make up paragraphs. Paragraphs make up columns of text and columns of text make up pages. When handling type we are dealing with designed letterforms. Only very few experienced typographers take the time and considerable effort to design new typefaces. Though over the years

Individual letters make words and words make up sentences (or lines) and lines make up paragraphs. Paragraphs make up columns of text and columns of text make up pages. When handling type we are dealing with designed letterforms. Only very few experienced typographers take the time and considerable effort to design new typefaces. Though over the

Quark Default H&J

	Min	Opt	Max
W	85	110	250
C	0	0	4
H ✓	ON		
T	0		

Quark Default H&J

	Min	Opt	Max
W	85	110	250
C	0	0	4
H	OFF		
T ✓	4		

Quark Default H&J

	Min	Opt	Max
W	85	110	250
C	0	0	4
H ✓	ON		
T	0		

Quark Default H&J

	Min	Opt	Max
W	85	110	250
C	0	0	4
H ✓	ON	HZ 14mm	
T	0		

Individual letters make words and words make up sentences (or lines) and lincs make up paragraphs. Paragraphs make up columns of text and columns of text make up pages. When handling type we are dealing with designed letterforms. Only very few experienced typographers take *the time and considerable effort to design new typefaces. Though over the years thou-*

Individual letters make words and words make up sentences (or lines) and lines make up paragraphs. Paragraphs make up columns of text and columns of text make up pages. When handling type we are dealing with designed letterforms. Only very few experienced typographers take *the time and considerable effort to design new typefaces. Though over the years*

Individual letters make words and words make up sentences (or lines) and lines make up paragraphs. Paragraphs makc up columns of text and columns of text make up pages. When handling type we are dealing with designed letterforms. Only very few experienced typographers take *the time and considerable effort to design new typefaces. Though over the years thou-*

Individual letters make words and words make up sentences (or lines) and lines make up paragraphs. Paragraphs make up columns of text and columns of text make up pages. When handling type we are dealing with designed letterforms. Only very few experienced typographers take *the time and considerable effort to design new typefaces. Though over the years*

User Defined H&J

	Min	Opt	Max
W	60	90	90
C	−5	0	10
H ✓	ON		
T	0		

User Defined H&J

	Min	Opt	Max
W	60	90	100
C	−3	0	5
H ✓	ON		
T ✓	2		

User Defined H&J

	Min	Opt	Max
W	85	100	250
C	0	0	0
H	OFF		
T	0		

Individual letters make words and words make up sentences (or lines) and lines make up paragraphs. Paragraphs make up columns of text and columns of text make up pages. When handling type we are dealing with designed letterforms. Only very few experienced typographers take the time and considerable effort to design new *typefaces. Though over the years thousands of typefaces have been*

Individual letters make words and words make up sentences (or lines) and lines make up paragraphs. Paragraphs make up columns of text and columns of text make up pages. When handling type we are dealing with designed letterforms. Only very few experienced typographers take the time and considerable effort to design new *typefaces. Though over the years thousands of typefaces*

Individual letters make words and words make up sentences (or lines) and lines make up paragraphs. Paragraphs make up columns of text and columns of text make up pages. When handling type we are dealing with designed letterforms. Only very few experienced typographers take the time and considerable effort to design new typefaces. Though over the years thousands of typefaces have been designed, so the choice for us is enormous. When we look at individual typefaces we can see that they each have their own particular characteristics and "personality." We may examine each of the 26 letters of the alphabet and can see *that their form and structure has a special relationship with the background on which it sits. Not only does the shape of the letter itself give it form, but the space around it accents its individuality.*

Rotis
−14

Rotis
−14 −2 −6

Rotis
−16 −2 −6

Rotis Sans, Semi Sans, and Serif showing overall tracking of −10 with letter pairs individually kerned as shown. Gray text indicates 0 tracking and no kerning of letter pairs.

SEMI SERIF 55 REGULAR
AaBbCcDdEeFfGgHhIiJjKkLlMmNnOoPpQqRrSsTt
UuVvWwXxYyZz1234567890;:'"!?@#£$&*()+=

SEMI SERIF 65 BOLD
AaBbCcDdEeFfGgHhIiJjKkLlMmNnOoPpQqRrSsTt
UuVvWwXxYyZz1234567890;:'"!?@#£$&*()+=

SERIF 56 ITALIC
*AaBbCcDdEeFfGgHhIiJjKkLlMmNnOoPpQqRrSsTt
UuVvWwXxYyZz1234567890;:'"!?@#£$&*()+=*

SERIF 65 BOLD
AaBbCcDdEeFfGgHhIiJjKkLlMmNnOoPpQqRrSsTt
UuVvWwXxYyZz1234567890;:'"!?@#£$&*()+=

ITC Stone Serif and Sans

ITC Stone Serif was designed as part of the extended Stone family and issued in 1987. This family includes three groups of typefaces: sans, serif, and informal in three weights of roman and italic. The three groups, designed to work with each other as well as on their own, were conceived and created by Sumner Stone, the distinguished American graphic and type designer and calligrapher. Each group has its own particular characteristics yet shares a basic constructional similarity: the proportions as well as the cap and x-heights are all the same. These built-in family characteristics are subliminal and so do not detract from the individuality of each style, but they do ensure a harmony and cohesion in the mixing of the Stone family, however flexibly used. The serif version has obvious thick and thin strokes but the overall effect provides the same optical color as the sans (compare this to Meta where the serif and sans versions share identical stroke widths, only the addition or subtraction of serifs distinguishing the two).

ITC Stone Serif was designed for the screen as well as paper and copes well with variances in screen resolution.

TYPE CHARACTERISTICS / SERIF

ITC Stone Serif has a chiseled quality, with crisp, angled serifs. In the capitals, the "K" and "R" both have trimmed lower diagonal strokes with slightly flared, rather than seriffed, endings. The capital "K" has a half serif on the upper diagonal stroke. The lowercase "J" has a short tail that falls below the baseline and ends with a sharp cut. The capital "Q" has a comparatively long, slightly curved tail with a minutely flared ending. The lowercase "g" is two-story both in the roman and italic versions and is unique in the family.

The italic has a written, pen-like quality with more apparent contrast in the stroke weights than in the roman. It has a distinctive script-like lowercase "f" with a short, serif-less tail and a crossbar at the top of the x-height line. The lowercase "a" has a slight squaring at the top and the tail of the lowercase "y" finishes with an outer half serif.

ABCDEFGHIJKLMNOPQRSTUVWXYZ
abcdefghijklmnopqrstuvwxyz1234567890;:'"!?@#£$&*()+=

Quark Default H&J	Min	Opt	Max
W ✎	85	110	250
C ✎	0	0	4
H ✓	ON		
T	0		

Individual letters make words and words make up sentences (or lines) and lines make up paragraphs. Paragraphs make up columns of text and columns of text make up pages. When handling type we are dealing with designed letterforms. Only very few experienced typographers take the time and considerable effort to design new typefaces. Though over the years thousands of typefaces have been designed, so the choice for us is enormous. When we look at individual typefaces we can see that they each have *their own particular characteristics and "personality." We may examine each of the 26 letters of the alphabet and can see that their form and structure has a special rela-*

User Defined H&J	Min	Opt	Max
W ✎	80	90	150
C ✎	0	3	5
H ✓	ON		
T ✓	3		

Individual letters make words and words make up sentences (or lines) and lines make up paragraphs. Paragraphs make up columns of text and columns of text make up pages. When handling type we are dealing with designed letterforms. Only very few experienced typographers take the time and considerable effort to design new typefaces. Though over the years thousands of typefaces have been designed, so the choice for us is enormous. When we look at individual typefaces we can see that *they each have their own particular characteristics and "personality." We may examine each of the 26 letters of the alphabet and can see that their form and structure*

User Defined H&J	Min	Opt	Max
W ✎	70	100	120
C ✎	–3	0	5
H ✓	ON		
T	0		

Individual letters make words and words make up sentences (or lines) and lines make up paragraphs. Paragraphs make up columns of text and columns of text make up pages. When handling type we are dealing with designed letterforms. Only very few experienced typographers take the time and considerable effort to design new typefaces. Though over the years thousands of typefaces have been designed, so the choice for us is enormous. When we look at individual typefaces we can see that they each have their *own particular characteristics and "personality." We may examine each of the 26 letters of the alphabet and can see that their form and structure has a special relationship with*

Quark Default H&J

	Min	Opt	Max
W	85	110	250
C	0	0	4
H	ON		
T	0		

Individual letters make words and words make up sentences (or lines) and lines make up paragraphs. Paragraphs make up columns of text and columns of text make up pages. When handling type we are dealing with designed letterforms. Only *very few experienced typographers take the time and considerable effort to design*

Quark Default H&J

	Min	Opt	Max
W	85	110	250
C	0	0	4
H	OFF		
T	0		

Individual letters make words and words make up sentences (or lines) and lines make up paragraphs. Paragraphs make up columns of text and columns of text make up pages. When handling type we are dealing with designed letterforms. Only *very few experienced typographers take the time and considerable effort to*

Quark Default H&J

	Min	Opt	Max
W	85	110	250
C	0	0	4
H	ON		
T	0		

Individual letters make words and words make up sentences (or lines) and lines make up paragraphs. Paragraphs make up columns of text and columns of text make up pages. When handling type we are dealing with designed letterforms. Only *very few experienced typographers take the time and considerable effort to design*

Quark Default H&J

	Min	Opt	Max
W	85	110	250
C	0	0	4
H	ON	HZ 14mm	
T	0		

Individual letters make words and words make up sentences (or lines) and lines make up paragraphs. Paragraphs make up columns of text and columns of text make up pages. When handling type we are dealing with designed letterforms. Only *very few experienced typographers take the time and considerable effort to*

User Defined H&J

	Min	Opt	Max
W	60	90	110
C	-5	-5	5
H	ON		
T	0		

Individual letters make words and words make up sentences (or lines) and lines make up paragraphs. Paragraphs make up columns of text and columns of text make up pages. When handling type we are dealing with designed letterforms. Only very few experienced typographers *take the time and considerable effort to design new typefaces. Though over the years thousands*

User Defined H&J

	Min	Opt	Max
W	60	90	90
C	-5	0	10
H	ON		
T	2		

Individual letters make words and words make up sentences (or lines) and lines make up paragraphs. Paragraphs make up columns of text and columns of text make up pages. When handling type we are dealing with designed letterforms. Only very few experienced *typographers take the time and considerable effort to design new typefaces. Though over the*

User Defined H&J

	Min	Opt	Max
W	85	100	250
C	0	0	0
H	OFF		
T	0		

Individual letters make words and words make up sentences (or lines) and lines make up paragraphs. Paragraphs make up columns of text and columns of text make up pages. When handling type we are dealing with designed letterforms. Only very few experienced typographers take the time and considerable effort to design new typefaces. Though over the years thousands of typefaces have been designed, so the choice for us is enormous. When we look at individual typefaces we can see that they each have their own particular characteristics and "personality." We may *examine each of the 26 letters of the alphabet and can see that their form and structure has a special relationship with the background on which it sits. Not only does the shape of the letter*

User Defined H&J

	Min	Opt	Max
W	60	100	120
C	-3	0	5
H	OFF		
T	0		

Individual letters make words and words make up sentences (or lines) and lines make up paragraphs. Paragraphs make up columns of text and columns of text make up pages. When handling type we are dealing with designed letterforms. Only very few experienced typographers take the time and considerable effort to design new typefaces. Though over the years thousands of typefaces have been designed, so the choice for us is enormous. When we look at individual typefaces we can see that they each have their own particular *characteristics and "personality." We may examine each of the 26 letters of the alphabet and can see that their form and structure has a special relationship with the*

ITALIC	*AaBbCcDdEeFfGgHhIiJjKkLlMmNnOoPpQqRrSsTt UuVvWwXxYyZz1234567890;:'"!?@#£$&*()+=*
SEMIBOLD	**AaBbCcDdEeFfGgHhIiJjKkLlMmNnOoPpQqRrSsTt UuVvWwXxYyZz1234567890;:'"!?@#£$&*()+=**
SEMIBOLD ITALIC	***AaBbCcDdEeFfGgHhIiJjKkLlMmNnOoPpQqRrSsTt UuVvWwXxYyZz1234567890;:'"!?@#£$&*()+=***
BOLD	**AaBbCcDdEeFfGgHhIiJjKkLlMmNnOoPpQqRrSsTt UuVvWwXxYyZz1234567890;:'"!?@#£$&*()+=**
BOLD ITALIC	***AaBbCcDdEeFfGgHhIiJjKkLlMmNnOoPpQqRrSsTt UuVvWwXxYyZz1234567890;:'"!?@#£$&*()+=***

Stone

ITC Stone Sans

TYPE CHARACTERISTICS / SANS

Stone Sans enjoys a subtle change in stroke weights. The lowercase "g" is single story with a very generous rounded tail in contrast to the minimalist lowercase "j."

    All strokes are terminated at right angles to the stroke direction, particularly noticeable in the legs of the capital "K" and "R" and most pronounced in the bold version of the face.

In Stone Serif the legs of both the "k" and the "r" (both upper- and lowercase) have a sweeping flared finish that adds vitality to the letterform. A similar vitality has been achieved in the Sans versions by finishing the strokes at an angle, which is equally effective. Stone's ampersand is stylish and, with its open upper portion, is well balanced.

Setting Tips

Stone's generous letterspacing provides an opportunity for a decrease in tracking but not for extra leading.

AaBbCc

ABCDEFGHIJKLMNOPQRSTUVWXYZ
abcdefghijklmnopqrstuvwxyz1234567890;:'""!?@#£$&*()+=

Quark Default H&J			
	Min	Opt	Max
W ✐	85	110	250
C ✐	0	0	4
H ✓	ON		
T	0		

Individual letters make words and words make up sentences (or lines) and lines make up paragraphs. Paragraphs make up columns of text and columns of text make up pages. When handling type we are dealing with designed letterforms. Only very few experienced typographers take the time and considerable effort to design new typefaces. Though over the years thousands of typefaces have been designed, so the choice for us is enormous. When we look at individual typefaces we can see that they each have their own par*ticular characteristics and "personality." We may examine each of the 26 letters of the alphabet and can see that their form and structure has a special relationship with the back-*

User Defined H&J			
	Min	Opt	Max
W ✐	80	90	150
C ✐	0	3	5
H ✓	ON		
T ✓	3		

Individual letters make words and words make up sentences (or lines) and lines make up paragraphs. Paragraphs make up columns of text and columns of text make up pages. When handling type we are dealing with designed letterforms. Only very few experienced typographers take the time and considerable effort to design new typefaces. Though over the years thousands of typefaces have been designed, so the choice for us is enormous. When we look at individual typefaces we can see that they each have their own *particular characteristics and "personality." We may examine each of the 26 letters of the alphabet and can see that their form and structure has a special relation-*

User Defined H&J			
	Min	Opt	Max
W ✐	70	100	120
C ✐	–3	0	5
H ✓	ON		
T	0		

Individual letters make words and words make up sentences (or lines) and lines make up paragraphs. Paragraphs make up columns of text and columns of text make up pages. When handling type we are dealing with designed letterforms. Only very few experienced typographers take the time and considerable effort to design new typefaces. Though over the years thousands of typefaces have been designed, so the choice for us is enormous. When we look at individual typefaces we can see that they each have their own particular characteris*t*ics and "personality." We *may examine each of the 26 letters of the alphabet and can see that their form and structure has a special relationship with the background on which it sits. Not only does the shape*

Quark Default H&J

	Min	Opt	Max
W ✐	85	110	250
C ✐	0	0	4
H ✓	ON		
T	0		

Individual letters make words and words make up sentences (or lines) and lines make up paragraphs. Paragraphs make up columns of text and columns of text make up pages. When handling type we are dealing with designed letterforms. Only very few experienced *typographers take the time and considerable effort to design new typefaces. Though*

Quark Default H&J

	Min	Opt	Max
W ✐	85	110	250
C ✐	0	0	4
H	OFF		
T	0		

Individual letters make words and words make up sentences (or lines) and lines make up paragraphs. Paragraphs make up columns of text and columns of text make up pages. When handling type we are dealing with designed letterforms. Only very few experienced *typographers take the time and considerable effort to design new typefaces. Though*

Quark Default H&J

	Min	Opt	Max
W ✐	85	110	250
C ✐	0	0	4
H ✓	ON		
T	0		

Individual letters make words and words make up sentences (or lines) and lines make up paragraphs. Paragraphs make up columns of text and columns of text make up pages. When handling type we are dealing with designed letterforms. Only very few experienced typographers *take the time and considerable effort to design new*

Quark Default H&J

	Min	Opt	Max
W ✐	85	110	250
C ✐	0	0	4
H ✓	ON	HZ 14mm	
T	0		

Individual letters make words and words make up sentences (or lines) and lines make up paragraphs. Paragraphs make up columns of text and columns of text make up pages. When handling type we are dealing with designed letterforms. Only very few *experienced typographers take the time and considerable effort to design new typefaces.*

User Defined H&J

	Min	Opt	Max
W ✐	70	110	250
C ✐	-3	5	10
H ✓	ON		
T	0		

Individual letters make words and words make up sentences (or lines) and lines make up paragraphs. Paragraphs make up columns of text and columns of text make up pages. When handling type we are dealing with designed letterforms. Only very few experienced typographers *take the time and considerable effort to design new typefaces. Though over the years thou-*

User Defined H&J

	Min	Opt	Max
W ✐	70	90	115
C ✐	-3	0	5
H ✓	ON		
T ✓	2		

Individual letters make words and words make up sentences (or lines) and lines make up paragraphs. Paragraphs make up columns of text and columns of text make up pages. When handling type we are dealing with designed letterforms. Only very few experienced typographers *take the time and considerable effort to design new typefaces. Though over the years thou-*

User Defined H&J

	Min	Opt	Max
W ✐	85	100	250
C ✐	0	0	0
H	OFF		
T	0		

Individual letters make words and words make up sentences (or lines) and lines make up paragraphs. Paragraphs make up columns of text and columns of text make up pages. When handling type we are dealing with designed letterforms. Only very few experienced typographers take the time and considerable effort to design new typefaces. Though over the years thousands of typefaces have been designed, so the choice for us is enormous. When we look at individual typefaces we can see that they each have their own particular characteristics and "personality." We may examine each of the 26 letters of *the alphabet and can see that their form and structure has a special relationship with the background on which it sits. Not only does the shape of the letter itself give it form, but the space around*

User Defined H&J

	Min	Opt	Max
W ✐	70	90	105
C ✐	-6	-3	5
H	OFF		
T	0		

Individual letters make words and words make up sentences (or lines) and lines make up paragraphs. Paragraphs make up columns of text and columns of text make up pages. When handling type we are dealing with designed letterforms. Only very few experienced typographers take the time and considerable effort to design new typefaces. Though over the years thousands of typefaces have been designed, so the choice for us is enormous. When we look at individual typefaces we can see that they each have their own particular characteristics and "personality." We may *examine each of the 26 letters of the alphabet and can see that their form and structure has a special relationship with the background on which it sits. Not only does the shape of the*

ITALIC	*AaBbCcDdEeFfGgHhIiJjKkLlMmNnOoPpQqRrSsTt UuVvWwXxYyZz1234567890;:'"!?@#£$&*()+=*
SEMIBOLD	**AaBbCcDdEeFfGgHhIiJjKkLlMmNnOoPpQqRrSsTt UuVvWwXxYyZz1234567890;:'"!?@#£$&*()+=**
SEMIBOLD ITALIC	***AaBbCcDdEeFfGgHhIiJjKkLlMmNnOoPpQqRrSsTt UuVvWwXxYyZz1234567890;:'"!?@#£$&*()+=***
BOLD	**AaBbCcDdEeFfGgHhIiJjKkLlMmNnOoPpQqRrSsTt UuVvWwXxYyZz1234567890;:'"!?@#£$&*()+=**
BOLD ITALIC	***AaBbCcDdEeFfGgHhIiJjKkLlMmNnOoPpQqRrSsTt UuVvWwXxYyZz1234567890;:'"!?@#£$&*()+=***

PART 3

Custom, Web & Display Types

Designing your own fonts

Designing your own fonts

It may seem surprising that, with so many fonts already in existence, new designs are constantly in demand. However, if you liken typeface design to other art forms—music, dance, poetry, painting, and sculpture—it will become apparent that new variants of old letterforms and alphabets, including complete new font designs, will always be forthcoming, influenced by changes in culture, fashion, and communication media.

Type enthusiasts will never grow tired of looking for new typefaces and finding new uses for old ones—and some may even try their hands at designing their own. Designing and producing your own typeface has been made easier by the availability of dedicated type-design software.

Ikarus is a sophisticated type-digitization program used by leading typeface designers. Several years ago it was available on all the major platforms but, at the time of writing, there was no up-to-date Macintosh version. There was, however, a beta version, so the final program was imminent.

Macromedia Fontographer and Pyrus FontLab are two well-known, user-friendly type design applications, both capable of designing custom types and outputting them in TrueType and PostScript formats.

In simple terms, type-design applications are vector-based drawing software that enables single outline shapes (letterforms) to be assigned to selected keystrokes, thereby creating custom fonts. If only a bitmap font needs to be created, you can also work in pixel mode.

Characters may be designed entirely within the software, or EPS files may be imported and their outlines used to modify and create further designs. Kerning tables from Adobe Font Metrics may also be imported and edited. It is also an option for images created in, for

This close-up of a glyph shows Bézier control points (the blue squares), guidelines, the baseline, and the side bearings. Clicking on a Bézier control point reveals handles that can be adjusted to edit the curve.

Macromedia's Fontographer tool box contains familiar drawing tools. Each character is shown in its own window and is associated with a keystroke.

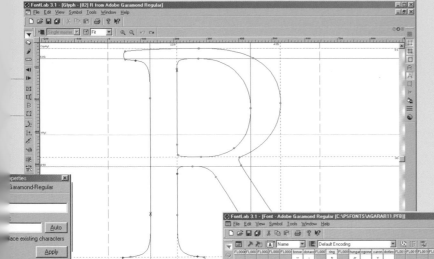

Three working windows in Pyrus FontLab. The foremost windows show the entire character set and the glyph properties, which let you select characters to edit and change details of their encoding. The rearmost window shows the glyph in editable outline form. The baseline, cap-height line, and side-bearings are all visible, along with guidelines to help in construction.

example, Macromedia Freehand or Adobe Illustrator to be pasted into the software and incorporated into the design of individual characters.

Old or forgotten typefaces are a rich source of material for scanning and can provide, by use of auto-trace tools, a character outline for further modification and change.

The clear interfaces of both Macromedia Fontographer and Pyrus FontLab are simple to understand and use. For example, Fontographer displays a window showing all characters in a font. Clicking one character in this window shows a close-up of the character for editing. The character can be viewed as an outline, a bitmap, and a filled-in letter, with tools for altering spacing and kerning. The outline view, with its accompanying toolbox, is the most important view initially, since it is in this window that the character is drawn and edited.

Anyone who has experience of drawing applications, such as Macromedia Freehand and Adobe Illustrator, will find Fontographer tools particularly easy to use. Bézier curves, corner and curve points, curve levers, and ways of combining shapes will all feel very familiar—as will other drawing tools, including the basic shape tools, the knife tool, and the zoom tool. A basic layer palette lets you show and hide guides, templates, and hinting information. All the usual transformation tools are featured, including scaling, skewing, and rotation. The outline view window has guides to help you to mark side bearings and character position. Imported fonts may be incorporated into other fonts by a process of interpolation, which blends one shape into another. Multiple Masters can be created in a similar way by morphing the extreme ends of design axes.

Tables of font metrics are viewable, as are kerning tables, which can be copied and, if required, modified. Bitmap fonts can be generated, and the bitmap window includes tools to help. However, now that outline information is commonly used to display type on screen, bitmaps will probably be of interest to only a specialized few.

Fontographer works on two levels: "easy" and "advanced." Much of what you need to do, by way of exporting characters to form a font or importing metrics, is semi-automated. Many dialog boxes have "easy-option" buttons to select, which is probably the best way to work, although there remains plenty of scope at the advanced level for those who enjoy digging deep.

WWW.URWPP.DE/ENGLISH/HOME.HTML

WWW.MACROMEDIA.COM/SOFTWARE/FONTOGRAPHER

WWW.PYRUS.COM/HTML/FONTLAB.HTML

Web and screen

D igital technology has raced ahead in giving us more ways of projecting our ideas and images to hungry audiences than just print on paper. It has opened up enormous opportunities for publishing data and creating inspiring interactive forms of communication. But, while those who work with print have been provided with new, exciting tools that enable them to hone their work to perfection, those who communicate through the screen have conversely had to suffer some serious drawbacks.

Screen-based products fall into two broad but merging categories: broadcast and online. Broadcast generally refers to television and video, while online refers to CD-ROM publishing and Web pages.

One problem that faces all designers of Web-based products is that of resolution and the display of fine detail. And fine detail is the key to quality in typographic presentation.

Online publications are quite likely to carry text-rich content, which may give rise to many problems associated with font fidelity and legibility of type at small sizes. The brightly colored pixels of the computer monitor are coarsely spaced, so they dramatically reduce the smoothness of image edge and fidelity of detail. The quality varies from one system to another: Microsoft Windows uses 96 ppi for the resolution of its display, whereas Apple Macintosh uses 72 ppi. You would thus expect type to look smaller on a PC but, since Windows uses a different method of displaying type, text sizes appear to be 3 to 4 points bigger on a PC than on a Macintosh screen.

In a CD-ROM publication the designer may impose total visual control: screens will always look the same to the viewer, regardless of the computer platform. Given the closed nature of a CD-ROM, specifying type of one's choice is not difficult, but choosing the right type for good screen viewing remains all-important.

The issue of type for the Web is more complex. The efficiency of a Web site depends heavily on the rapid transmission of very small files, which are

The screen grabs below show the same Web page on a PC and Mac screen. Although the images (GIFS and J-PEGS) are the same size, the text on the PC screen is larger than on the Mac screen. This is due to the difference in pixels per inch for the screen resolution.

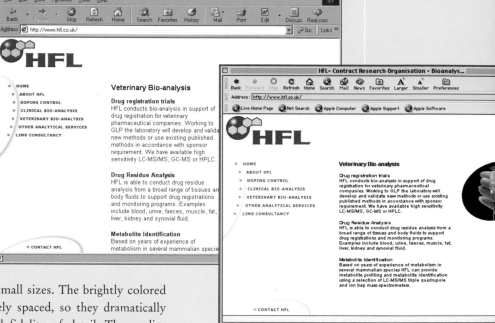

Above is a simulation of the difference in text size on a PC and Mac screen. At 72 ppi, the Mac screen resolution is 75% of the PC screen's 96 ppi.

then interpreted at the viewing end by a browser. This is where a whole series of problems can begin to create inconsistencies in the display of type and text on screen.

Web-page display depends on many variables, including monitor size, screen resolution, platform, browser capabilities, and each user's set-up. All these variables add to the uncertainty of how a page may ultimately display, and they make it hard for the designer to achieve fidelity and accuracy in Web layout and presentation. Different browsers impose different limitations on available fonts and sizes, and—to make matters worse—Windows and Macintosh have mostly different preinstalled fonts. Thus, one person's view of a page may differ from another's.

A Web-site designer can specify a font to be used but, if the font is not on the user's system, installed fonts must be used instead. As a result, font choices are often specified as "preferred, alternative, last resort," e.g., "Arial, Helvetica, sans serif." The obvious solution of serving the font with the page would add to download times and would risk infringing copyright in the font.

As a consequence, designers must complicate their task by specifying alternative fonts and sizes for different platforms and browsers, i.e., they must have enough knowledge to be able to anticipate what most users' systems will be capable of, and must then cover all eventualities. Even then, differences in monitor resolution and settings may alter how text is viewed. Hopefully, future browser and operating system releases will be shipped with larger sets of compatible fonts.

However, as problems increase, so do possible solutions. The following techniques and technologies are developing rapidly and will interest those who care about the quality of type.

Images on the left showing the default, largest, and smallest text alternatives in the Internet Explorer 5 browser.

CASCADING STYLE SHEETS

Cascading Style Sheets (CSS) are to Web pages what document style sheets are to DTP. Instead of tagging each element on a page with font, size, style, and color, the designer tags it with a style name. Each style name is held in one or more style sheets and the details are applied as the page is served. As for DTP, using CSS allows the appearance of pages to be changed globally by changing the style sheet, rather than each page. One immediate benefit is

that different style sheets can be applied in different situations, if a test that queries the user's platform and browser is included in the invisible section of the Web page. A PC user can be served with a different sheet from a Macintosh user and a Navigator user can be served with a different sheet from an Explorer user. Style sheets can thus take account of differences in platform and of how browsers interpret the styles. In theory, they are also easy to update when browsers or platforms get new capabilities, though such developments move slowly. Use of CSS thus copes efficiently with client-side (user) variations, but does not guarantee the outcome.

FONT EMBEDDING—FONT OBJECTS

As Web pages grow more complex, font fidelity becomes a more critical issue, impelling software companies not only to increase their user-base but also to establish leadership in Web-font technology.

The Web-page designer's aim is to make type appear exactly as intended on anyone's screen. Sending font files with Web pages would seem the obvious solution, but it is slow and rarely legal. Some software utilities can overcome this by capturing the shape of characters (glyphs) used on a page and sending only those used in rasterized form (font objects), incorporating hinting and antialiasing for optimal clarity. The file is smaller than a complete outline font file, and can easily be served along with other files and linked to pages by HTML code. On the client side, browser software can interpret these characters for display.

WEFT

Microsoft has developed a font-embedding utility called WEFT (currently at version 3, thus WEFT 3) that lets designers link "font objects" to Web sites. The objects are readable by Microsoft Explorer but not yet for Netscape Navigator and the Macintosh platform; however, users have reported that WEFT 3 works under Apple operating systems running the latest versions of some PC emulations. Microsoft's long-term plan is to build WEFT technology into both leading Macintosh and Windows software.

TrueDoc and WebFont Maker

TrueDoc is a technology developed by Bitstream for software developers to incorporate into their products. Bitstream also sells this technology in WebFont Maker, a product that lets you make any TrueType or PostScript font on your system into a "dynamic font" for use on Web pages. Technology for reading TrueDoc information is built into the latest versions of Netscape Navigator, and there is a plug-in for Microsoft Internet Explorer for Windows. A similar plug-in for Microsoft Internet Explorer on the Macintosh is being developed.

The dialog box for defining style sheets for Web pages. CSS allows the appearance of a Web page to be changed by changing the style sheet, rather than the page.

All the text and ornaments here are fonts—no images are used at all! Unusual typefaces would normally be treated as static graphics. From http://www.microsoft.com/typography/web/embedding/demos/3/demo3.htm

When a font chosen by a Web-page designer is not installed on the viewers' system, the font can be seen using a plug-in like Webfontmaker.

WEB FONTS

Alongside the desire for certainty that your chosen typefaces will be displayed correctly on the client side, there is another highly significant issue—that of rasterizing characters to be served with a Web page. Rasterizing should ideally ensure that antialiasing is used to help smooth the edges of the type, thereby improving its appearance when rendered on screen.

Rasterizing characters into "font objects" may not be appropriate for all situations, and some font manufacturers have done important work in selecting, modifying, and designing fonts that have built-in features that allow them to look good on screen.

Monotype ESQ

Agfa Monotype has produced a library of more than 150 high-quality fonts that have been specially designed and fine-tuned with screen resolution in mind. They are a special kind of TrueType font, called ESQ (Enhanced Screen Quality), and they perform well at relatively small sizes without font embedding.

TrueType Core Fonts for the Web

Microsoft has optimized a number of TrueType fonts to work well on the Web. They include the full Windows Glyph List 4 (WGL4) and have greatly extended character sets. Users can download the fonts free from Microsoft. Although different versions are used on different platforms, they support cross-platform compatibility and, most significantly, the Times New Roman and Courier New core fonts will look the same size when viewed on either platform.

The product developers hope that OpenType fonts will quickly replace PostScript and TrueType fonts. OpenType's extended character set should ensure that special characters can be displayed properly on all platforms and in all software used on the Web—a task in which the older font formats currently fail in Web work.

Given the uncertain development and compatibility problems that surround different embedding techniques, browsers, and monitor displays, it is still a tricky task to achieve font fidelity on the Web. Complete solutions that let anyone,

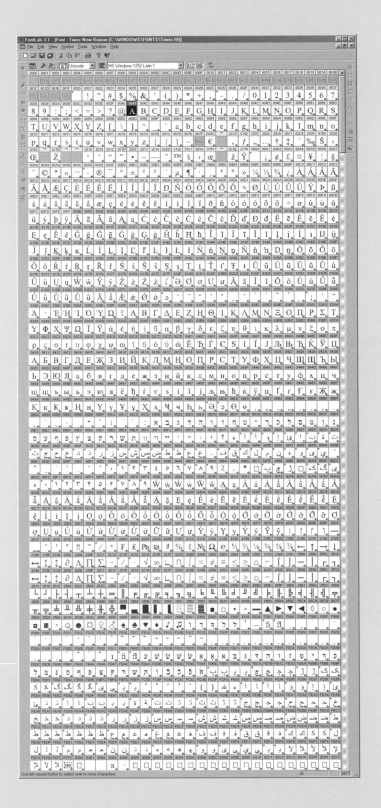

The entire character set of Times New Roman, a Microsoft TrueType core font. The file is large because it contains the full Windows Glyph List 4 (WGL4) set of 652 characters, covering multiple writing systems including Western, Central, and Eastern European, as well as characters required in Greek and Turkish.

Fonts that are not likely to be found on the user's system can be used to create type as a graphic by using a bitmap image editor such as PhotoShop. The image may be exported as a GIF file, which provides a transparent background, or as a J-PEG file. Image files must be kept as small as possible.

anywhere, view Web pages as you designed them are still some way off. Currently, designers accept that total control over how their Web products are viewed is not always achievable, but they use every measure open to them to reduce potential problems. With careful planning and selection, you can make sure that at least most of your target audience will be able to view a true representation of your product. Until recently, if you were restricted to choosing system fonts, typographic decision-making was relatively easy: you would choose either a serif or sans serif typeface in the knowledge that Macintosh users would view your text in Times or Helvetica, and that PC users would view the same information in Times New Roman or Arial. Font choice is, however, now being increased. Take care in your selection of fonts, look out for the developing font-embedding software tools, and take advantage of font formats designed specifically for the Web.

TYPE AS A GRAPHIC

If you wish to use a special typeface that users are unlikely to have installed, you can bypass entirely the difficulties of font formats and embedding tools by converting the text to a "picture." This will ensure that the user views your type exactly as you intend. However, the type is fixed—it will not flow as "live" text does—and uses much more memory. Text as a graphic therefore works best for pieces of display type. You can create such graphics in a bitmap image editor, e.g., a picture editor such as Adobe PhotoShop, which will let you create, arrange, color, and antialias the type, and

perhaps to apply suitable special effects. Create the graphic at the intended viewing size and at 96 ppi; enlarging it later will greatly reduce its quality. Save the file in GIF or J-PEG format and import it to the Web page. Good quality type may also be exported as a PDF page which will preserve the typeface and layout faithfully.

TAKING POSITIVE STEPS TO HELP TYPE LOOK GOOD

Even if you are able to use a font-embedding utility or to create type as a "graphic," you will still have to exercise care in your font selection. As we have seen, choice of type size is a particularly critical factor in Web-page design, since low resolution greatly restricts the effective display of small type. It is easy to see that a letter that is well rendered in a 25 pixel x 25 pixel grid will lose much detail if rendered in a 7 pixel x 7 pixel square. As the number of available pixels decreases, so does the screen's ability to show detail. Aim to find fonts and choose sizes where pixels can at least match up to the true outline shape and spacing. For example, it would invite trouble if you chose a typeface and size where the

Microsoft's core system fonts, which are available on both Macintosh and PC platforms, now allow a greater, albeit still narrow, range of fonts to be specified in browsers for Web page display.

thickness of a stroke or a character space was less than a pixel. Fine details, such as serifs and cross-bars, could easily, at small sizes, be less than a pixel; that particular detail would thus be lost. Similarly, the positioning of character strokes may not coincide with the pixel grid, which will corrupt stroke widths and spacing. And although hinting and antialiasing techniques can help to create the illusion of smoothness and to minimize jaggedness and distortion, the best results will always be achieved by hinting and antialiasing types that are strong and suitably robust in detail to begin with.

MICROSOFT READER

Great strides have been made by all sections of the communications industry to improve text display on computer monitors, and thus to improve Web-page display. But the Web is not the only area of digital expansion that involves the use of type; there is also the rapidly expanding market of pocket PCs. The pocket PC's portability and LCD display have made it an ideal platform for the development of the e-Book. Of interest to publishers and book designers is Microsoft's ClearType technology, demonstrated in the first e-Book, which was previewed in New York in April 2000. ClearType technology takes advantage of the rather high resolution of LCD displays and uses hinting and antialiasing to produce smoother characters.

ClearType forms the basis of the Microsoft Reader, Microsoft's e-Book reading software. E-Book software can closely follow the way printed books behave by emulating page-turning and providing bookmarks, annotation, and indexing. The e-Book will have a significant effect on portable computing, as dozens of book titles can be downloaded to portable devices, which may also be capable of a whole range of other digital forms of data manipulation and communication. Where there are new media for visual communication, there will also be creative challenges for the designer, publisher, and typographer. Until recently, screen-based reading has, of necessity, been delivered in bite-sized chunks, requiring a marked change of approach by designers previously used to designing for print. The advent of the e-Book, with all its implications, will encourage a reassessment—with a difference—of book layout and typographic skills.

Font Smoothing

The process of antialiasing in ClearType font smoothing is more intelligent than bitmap image software's antialiasing, evenly dispersing intermediate pixels between two colors, simulating a smooth edge. Font smoothing technology currently places intermediate gray pixels strategically to lend clarity to the essential structure of the letterform.

HTML

To the disappointment of many graphic designers who have become accustomed to the precision and ease of positioning type and graphic elements using layout programs, HTML (hypertext mark-up language)—which has become the Internet—uses seemingly clumsy sequential placement rules. The original intention of HTML's early developers was to provide a fast, effective way of transmitting simply structured documents. Layout was of secondary concern. Today's accelerating demand for image- and type-rich content on the Web was not envisaged in HTML's early days.

HOW DOES HTML WORK?

HTML is a set of directions for assembling picture and text files with layout and color information for transmission as encoded alphanumeric characters through computer networks and telephone lines. Web-browsing software reassembles the files according to the HTML instructions to produce a display on the end-user's monitor: Netscape Navigator and Internet Explorer are the best-known browsers. Most users connect to the Web through telephone line. Many lines are fiber-optic and allow large amounts of digital data to be sent very quickly, but millions of lines are still copper wires, which are both slower and less secure. Developments like ADSL (Asymetric Digital Subscriber Line) have increased transmission speeds but browsers will still receive and display smaller files faster than larger ones. An efficient Web site is one that uses the smallest files consistent with obtaining the desired results.

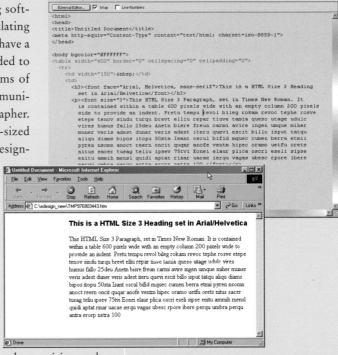

The upper window shows HTML written as plain text following set conventions to provide layout and rudimentary typographic detailing. The lower window demonstrates how a browser interprets that text to display a Web page.

Emigre fonts

Emigre fonts

Left: The first issue of *Emigre*: The Magazine that Ignores Boundaries.

Below: An Emigre font catalog.

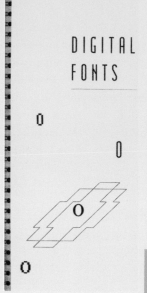

Emigre Graphics began as a small design studio founded in the 1980s by Rudy VanderLans, a young Dutch designer living in California, and Zuzana Licko, his Czech wife. Licko had moved to California in 1968, at the age of seven. VanderLans had emigrated to the United States in 1981, after receiving a conventional design education in Holland, and had studied photography at the University of California at Berkeley.

VanderLans became fascinated by a group of Dutch artists who had formed a collective called Hard Werken. This group had an unconventional view of design, breaking away from Swiss functionalism. Strongly influenced by Hard Werken's expressive, unstructured ways of working, VanderLans published the first copy of a cultural magazine, *Emigre,* which featured experimental work by poets, architects, painters, and artist friends.

Significantly, this new publishing venture coincided with the launch of the Macintosh computer. Despite the power and opportunities it put in the hands of individuals, the initial reaction of many was that the Macintosh's small, coarse monitor and dot-matrix print output were raw and crude, and it attracted much severe criticism. VanderLans and Licko were invited to test the new personal computer, and they immediately saw the possibilities of using it to help generate artwork. Licko had gained experience of computers from helping her father with them, and she had experimented, during a Berkeley computer programing class, with designing low-resolution type.

In different ways, VanderLans and Licko grasped the unique opportunities offered by the Macintosh. VanderLans used MacPaint software for illustrations, and Licko began to use font-editing software to create low-resolution custom fonts to store and use on the Mac. She was fascinated by the crudity and so-called limitations of the new technology and, rather than fight it, she worked innovatively with it.

Licko's early fonts were used to publish subsequent issues of *Emigre* as it began to gain a reputation among

Emigre magazine, published by Rudy VanderLans, using Licko's early bitmap fonts. Shown here, the cover of Issue 14 with the word "heritage" set in a low-resolution fonts.

The birth of the user-friendly personal computer. Rudy VanderLans's apt illustration of the Apple Macintosh.

Early Emigre typefaces from around 1985 (much enlarged)
Oakland 6
Emperor 15
Emigre 14

designers as a designer's magazine—an unconventional publication that broke with traditional typographic and layout concepts. Driven by small budgets, the magazine refused the restrictions of the accepted norms of reprographic origination and printing. The typewriter, Xeroxed images, and then the Macintosh provided much of the magazine's artwork. Licko continued to design and produce low-resolution fonts, and she was soon designing fonts for Adobe Systems. Further editions of the magazine featured an increasing number of new, innovative fonts that—together with the magazine's sharp and inventive design and layout—challenged many traditions and conventions in contemporary design through unashamed vitality and experimentation. Emigre fonts and the *Emigre* magazine significantly influenced artists and designers all over the world and became a byword for cutting-edge graphic design.

The personal computer—primarily the Macintosh—swiftly brought graphic designers, artists, and, in particular, typographers unprecedented freedom. Unfettered by specialized typesetters and a history of limited font foundries, anyone with a personal computer on their desk could either design their own type or select typefaces from a suddenly new, exciting range of fonts.

Emigre fonts are marked by a stark but disarming originality that takes its provenance from the birth of digital imaging. Uncompromisingly coarse pixel resolutions drove the design concepts and shaped the letterforms. The thread that runs through most of Licko's spirited typeface designs is economy—of space and digital storage. This is most evident in the pixelated simplicity of her earliest work and the "memory-light" faces, Matrix Book and Variex, which have vector outlines using a minimal number of Bézier control points and curves, producing a clarity of form that owes little to traditional, pen-derived letterforms.

Emigre fonts and *Emigre* magazine truly show that a direct engagement with new technology can produce groundbreaking results, and demonstrate clearly that innovation and creativity can be inspired, rather than stifled, by perceived limitations in the medium.

Although current font technology helps to smooth and simulate the crispness of traditional type, it is interesting to note that we still have freedom of choice to select screen-friendly fonts, designed or adapted for low-resolution pixelated displays. These type forms, although less familiar and quirky, may yet, through frequent use, become as familiar as print influenced fonts are. In answer to criticisms that her fonts were illegible, Zuzana Licko commented, "You read best what you read most" (*Emigre: Graphic Design into the Digital Realm*. Rudy VanderLans *et al.,* John Wiley and Sons, 1997).

Many years on from Emigre's modest beginnings, VanderLans and Licko now head their own company, Emigre Inc., a publishing house and digital type foundry based in Sacramento, California. The Emigre font list runs to some 50 families, and Emigre typefaces continue to contribute verve and energy to type design and type use.

WWW.EMIGRE.COM

Today's major players

Today's major players

The worlds of printing and type creation were turned upside down, shaken and supercharged when typesetting, layout, and page make-up went digital. An explosion occurred in the type technology and type design industry, and the amount of entrepreneurial activity in this area became astounding, as opportunities in type design and font marketing grew. The information and communications revolution affected millions of peoples' lives, so the stakes for those investing in core software activities were extremely high. Whoever was able to secure the largest user base and dominate standards of interconnectivity and display would inevitably reap the greatest rewards in the vast new market.

Long-established companies with traditional backgrounds in the graphic arts business were quick to buy software expertise and form alliances in order to get their share of the new market-place. Small software companies devising robust technology flourished in this competitive and often ruthless field. Less aggressive but just as vigorous was the growth of supply and support enterprises, including a growing number of font libraries and collections offered to the design pro-fession. Software became available for digital type design, enabling many more designers to enter the field of type design as a specialty.

FONT DEVELOPMENT, FONT LIBRARIES, AND TYPEFACE DESIGNERS

Who does what, and to what extent, in the field of type design and marketing is confusing, because there is much overlapping activity in the field. The

Thanks to the Internet it is easy to see who is doing what in the field of typeface design. Dozens of independent font foundries now have their own sites alongside the larger corporations. FontShop has an elegant and informative site and Microsoft's Web pages on typography are mines of information for the technically minded.

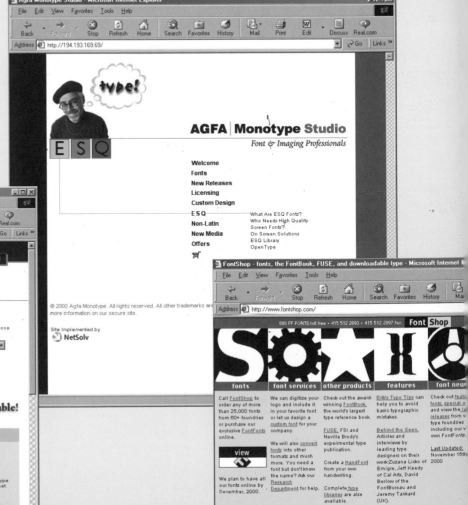

main activities are typeface design, font engineering, and the marketing of font collections. There are independent type designers who sell their designs to other concerns, which then market them in font libraries. There are companies that design and market their own fonts, and also license their products to other libraries. There are large companies that offer extensive libraries of fonts, and also commission new typeface designs to increase and diversify their range. Finally, there are companies that are primarily involved in the technology of font-handling, digitization, the mechanics of spacing and font metrics, the creation of font formats, and the establishment of their standards within the industry. Yet they, too, commission type designs and publish their own extensive font libraries.

The fact that many companies involved in the commissioning and sale of fonts also market many of their competitors' font collections, or include numerous fonts from other libraries in their own collections, makes for a blurring of roles that is not easy to define.

Adobe Systems
Adobe Systems, the creator of PostScript (the universally adopted page description language that translates vector outlines to bitmaps) has grown into one of the world's leading graphic arts software companies. Based in California, it markets an impressive range of graphics programs, the most notable of which are Adobe Acrobat, Adobe PhotoShop, Adobe Illustrator, Adobe GoLive, and Adobe InDesign.

Adobe developed the PostScript Type 1 font format and has collaborated with Microsoft to introduce the OpenType font format that will ultimately, it is expected, supersede Type 1 bringing with it cross-platform compatibility.

Adobe has its own extensive type library of 2500 typefaces by well-known type designers and foundries. Adobe fonts are also found in many other large collections and libraries.

WWW.ADOBE.COM/TYPE

Agfa Monotype

Monotype, based in Redhill, England, was one of the best-respected companies in type composing, with a 100-year history in hot-metal type-casting equipment. Its name was synonymous with high-quality type design and innovation. Agfa-Gevaert, a German-Belgian company with origins in photographic films, chemicals, and processes, merged in 1989 with a successful U.S. phototypesetting company, Compugraphic, to form the Agfa Corporation. In 1997 Agfa acquired Monotype Typography, as it was then known, and two years later the Agfa Monotype Corporation was formally launched. In 2000 Agfa Monotype acquired ITC (International Typeface Corporation), forming an impressive grouping of font-design and imaging technologies and becoming an even stronger player in the development of digital software for the communications industry. The company's WorldType technology led the way in using the Unicode 2 character-encoding standard, which is capable of supporting more than 49,000 characters.

The Agfa Monotype library of fonts is one of the largest collections, containing some 8000 fonts, including the Monotype Classics Collection and Creative Alliance exclusives, alongside types from Adobe, ITC and others.

WWW.MONOTYPE.COM

Bitstream

Bitstream, set up in 1981, was the first independent digital typefoundry. It has built up an impressive library of almost 2000 typefaces from "Bitstream Originals"—typefaces designed in-house or specially for the company—to many other sources, including ITC and leading typeface designers. Following Adobe's publication of the specification, Bitstream was the first company to release Type 1 fonts in great quantity and was noted for offering its complete range of fonts in both Type 1 and TrueType formats.

All Bitstream fonts, from whichever designer or collection, are redigitized, engineered, and hinted to Bitstream standards. Though best-known for its library and as a supplier of fonts, Bitstream has significant expertise in font technology and its development of TrueDoc imaging (discussed on page 164) provides solutions for font rendering and adding dynamic fonts to Web pages.

WWW.BITSTREAM.COM

WWW.MYFONTS.COM

Microsoft

Microsoft, the software giant, has also had a fundamental and significant involvement in a range of font technologies. Microsoft collaborated with Apple computers to create the TrueType format, and has recently worked with Adobe to develop OpenType or TrueType Open. The Microsoft Reader, the software behind e-Book development, is also of great interest: combined with Microsoft's ClearType, it should have an enormous impact on mobile computing, providing many graphic and typographic opportunities for e-Book designers and publishers.

U&lc, ITC's typography periodical was a radical showcase for many new and revival fonts. Produced in a large newsprint format, it became synonymous with exciting, new, and innovative typography. Regrettably, for some of us, the format has been reduced to "small magazine" size, as shown here.

Microsoft's TrueType core fonts—although the range is, as yet, modest—will be useful to Web-site designers, since they should provide better compatibility across platforms and among competing browsers.

In typeface marketing, there is currently no large library of Microsoft typefaces, although the company has employed well-known, highly respected designers to create some of its core system fonts. However, Microsoft publishes comprehensive and detailed technical information on many aspects of type design and typography.

WWW.MICROSOFT.COM/TYPOGRAPHY/ABOUT.HTM

TYPE COLLECTIONS AND LIBRARIES

Linotype Library GmbH

The lineage of the Linotype name stretches back to 1886, when the first hot-metal type casting machine, invented by Ottmar Mergenthaler, was installed in New York. After a long and varied collaboration with the Stempel foundry, and playing a distinguished role in the phototypesetting era—with well-remembered typesetters including the Linotronic 300—Linotype began, in 1987, to produce PostScript typefaces. Before this period of change, Linotype fonts had been available only on the company's own typesetting equipment, but a licensing agreement with Adobe has now made Linotype PostScript fonts for imagesetting widely available.

Now a subsidiary of Heidelberger Druckmaschinen AG, it operates as Linotype Library GmbH and is known as one of the larger libraries and suppliers of fonts.

WWW.LINOTYPELIBRARY.COM

ITC—International Typeface Corporation

The International Typeface Corporation, better known as ITC, was formed in 1969, in the phototypesetting years, and was acquired by Agfa Monotype in 2000. ITC, as a company that supplied and marketed type designs but never actually made fonts, was the concept of the renowned American typographers Herb Lubalin and Aaron Burns, in conjunction with Ed Rondthaler of Photo-Lettering Inc. Many designers and typographers felt ITC's impact through *U&lc* (Upper- and lowercase), its powerful and unusually large format publication. *U&lc* showcased classic type revivals as well as new type designs, and featured work by both established and up-and-coming type designers. ITC has been a significant force in the development of type and design protection. Its stock of more than 1000 designs now includes the Fontek collection, developed by Letraset, originator of the dry-transfer lettering system. *U&lc* magazine is still printed and published, although in smaller format, and there is an online version.

WWW.UANDLC.COM

Berthold

Berthold has a history going back almost 150 years and is among the globally significant type libraries. Founded as a brass-rule factory by Hermann Berthold, acquisitions made it the largest typefoundry in the world soon after World War I. Like all the leading type foundries, it moved into photocomposition in the 1950s, continuing to develop its

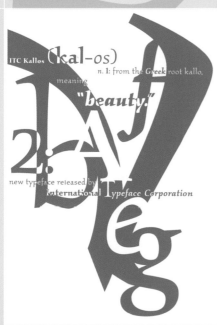

Shown below is the Grimshaw font, Kallos, displayed on a typical page from *U&lc*, making full use of the old generous format.

outstanding position in the field of type composition. Berthold's photocomposition systems were considered to represent the very best in type imaging. The company also developed the Berthold Exklusiv Collection, a collection of typefaces by highly respected type designers. Between 1960 and 1990, renowned type designer Günter Gerhard Lange was the inspiring and perfectionist Artistic Director in charge of all typeface development.

Now based in Chicago, Illinois, Berthold continues to offer the Berthold BQ Exklusiv Collection as well as the Adobe Berthold Exklusiv Collection.

WWW.BERTHOLDTYPES.COM

FontShop International

Founded in the 1990s, and based on the ideas of the prestigious and influential German type designer and typographer Erik Spiekermann, FontShop International (FSI) enjoys a high profile among font libraries. The excellent presentation and comprehensive content of its famous yellow type catalog, the *FontBook,* with its unique divider pages, gives it an unmistakable identity. FSI is a network of outlets in Europe and North America, all of which stock the contents of the catalog—an extensive and diverse collection of typefaces from a wide range of font developers and designers. FSI's own independent label, FontFont, represents and supports new designs from digital type designers.

FSI's standing has been enhanced by its exclusive, internationally acclaimed, multimedia publication, *FUSE.* Devoted to experimental typography, *FUSE* was conceived by Neville Brody, one of the most influential type designers of the digital age.

WWW.FONTSHOP.COM

Creative Alliance

The Creative Alliance library is a collection of typefaces assembled by Agfa Monotype from a host of other collections to form possibly the biggest type library in existence. Approximately 7000 fonts make up the library, which includes collections from ITC, Adobe, Agfa Monotype, and many independent type foundries and type designers.

All available on one CD-ROM, the Creative Alliance library contains PostScript and TrueType fonts and more than 150 ESQ (Enhanced Screen Quality) fonts that improve on-screen readability.

WWW.AGFAMONOTYPE.COM

TYPE DESIGNERS

Since the emergence of the personal computer, an increasing number of typographers have worked in the field of typeface design. Many are producing stylish and beautiful new faces or reviving classic designs. Others are involved in designing more transitory outrageous or fashionable letterforms, and some are producing innovative and lasting working tools for today's graphic designers and their successors. The quality of work by many of these type designers is often exceptionally high and, in time, some of their

Font catalogs show extensive collections and are generally well indexed. Though listed in alphabetical order, some typefaces may be listed under aliases or new names. When purchasing collections, make sure you have fully understood which and how many fonts make up that particular collection.

designs may prove to be new classics in their own right. But, given the constant development of communications technology, the notion of classic letterforms may well come to be radically redefined. At present, however, type's rich heritage continues to exert considerable influence on the typeface designer. The type "gurus" of today are, with some notable exceptions, those who have traveled the entire journey from metal type, through film, to digital type imaging. Here are just a few of the great contributors who have added to the richness of today's typeface design and typography, piqued or satisfied our curiosity about letters, and increased our understanding of the art and craft of visible speech.

abcdefg hijklmn opqrstu vwxyz

Matthew Carter

British type designer Matthew Carter, whose father, Harry, was an eminent type historian, has made an enormous contribution to the art and science of type design. His long career, both in Britain and in the United States, has embraced the worlds of metal type, photosetting, and digital type imaging. Trained as a punch cutter in the Netherlands, Carter became responsible for Crossfield's typographic program in the 1960s, and in 1965 became the in-house designer for Merganthaler Linotype in New York. In 1971, he returned to England to work for Linotype and in 1980 became consultant to HMSO (Her Majesty's Stationery Office). The following year, together with Mike Parker, he set up Bitstream, the first totally digital, and independent, type foundry. He is now principal of Carter and Cone Type based in Cambridge, Massachusetts, which he formed with Cherie Cone in 1991.

Matthew Carter, type designer and cofounder of Bitstream, responsible for such classic fonts as ITC Galliard, Verdana, Georgia, and Bell Centennial, among many others.

The list of Matthew Carter's type designs is long and impressive. It includes sole or joint responsibility for the well-known fonts Bitstream Bell Centennial, and Charter, ITC Galliard, Skia, and Sophia, and many others.

He was recently commissioned by Microsoft to design several good, readable screen fonts. Owing to the tremendous strides made in the development of screen-based products, Web-page applications, and the emerging e-Book technology, Microsoft had recognized that access to quality typefaces that read well on screen was becoming a pressing matter for authors and designers. The resulting typeface designs were Verdana and Georgia, both now included in Microsoft's range of core fonts.

Carter is noted for his long-established ability to design type to suit the requirements of constantly changing imaging media. The excellence of his work and his outstanding skill in designing type for, and matching it to, specific contexts have rightly received international recognition.

Adrian Frutiger

Swiss-born Adrian Frutiger has had a far-reaching influence in the graphic arts. Initially apprenticed to a printing company, he later graduated from the Zurich School of Arts and Crafts, and then took up a post in Paris with Deberny & Peignot, the French type foundry. Frutiger has designed numerous successful typefaces for metal, film, and digital media. He is probably most famous for his highly successful typeface, Univers, the design of which broke new ground in type design. Designed both for metal and film composition, Univers was an ingeniously conceived family of 21 weights and

ITC GALLIARD BLACK ITALIC

Signage at Charles de Gaulle airport, Roissy, designed by Adrian Frutiger. Letters are about 10 cm in height so they can be seen from a distance of 20 meters.

Adrian Frutiger—probably best known for Univers, for which he designed 21 weights and widths (numerically identified), thus developing the concept of variants derived from a single style (as now exemplified by Multiple Master fonts).

EXTENDED
CONDENSED
EXTENDED BLACK
LIGHT
REGULAR
UNIVERS

widths, each identified by a unique number assigned from a matrix of qualities. This comprehensive type family, with its inherent rationality and apparent simplicity of form, provided designers with an extraordinarily flexible set of typographic tools. It is highly probable that Univers was one inspiration behind the development of today's Multiple Master fonts. Frutiger also worked extensively on alphabets for signage systems. The Frutiger typeface was adapted by Linotype from a signage design, Roissy, which was used throughout the Charles de Gaulle airport in Paris. It is noted for its clarity and elegance. Adrian Frutiger's work is a benchmark of excellence in typeface design.

Günter Gerhard Lange

H Berthold AG, Germany's principal type foundry in the second half of the 20th century, owes much of its reputation for typographic excellence to the work of its Artistic Director, the legendary Günter Gerhard Lange. Lange studied calligraphy, printing, and typography, and initially worked mainly as a freelance graphic artist. In 1950 he moved into the field of type design, and from 1960, when he was appointed artistic director to Berthold, the development of typeface designs for the Berthold Exklusiv Collection was firmly under his creative control. In 1971, Lange was appointed to the Berthold board.

Under the influence, exacting standards and meticulous care of Lange, Berthold's phototypesetting systems and the quality of their fonts became synonymous with the typographic ideals of precision and distinction. Lange contributed to a growing and respected library of fonts by designing many classic typeface revivals, including Berthold Garamond, Baskerville Book, Caslon, Bodoni Old Face, and Walbaum Book, as well as several new type designs, including Boulevard,

Hermann Zapf, whose creative skills as a type designer are underpinned by his love of calligraphy, as is evident in many of his classic type designs.

ZAPF DINGBATS

Günter Gerhard Lange was responsible for many Berthold classic typeface revivals and for such new faces as Boulevard, Champion, and Arena New.

BOULEVARD

Champion, and Arena New. While working for Berthold, Lange also taught graphic design and typography at colleges in Germany.

Although he "retired" in 1990, Lange has continued to lecture and teach. His extraordinary skills and immense standing in the typographic field brought him an almost unprecedented invitation to return to Berthold as artistic consultant, 10 years into his "retirement."

Hermann Zapf

Some of the greatest typefaces of the 20th century were designed by Hermann Zapf, a prolific type designer ranking high among the great names in typographic history.

Zapf was born in Nuremberg, Germany, and was apprenticed as a retoucher in a printing company. In 1938, he went to work in Frankfurt, where he learned about printing, punch-cutting, and other related skills. At the end of World War II, he took up the post of design director at the D Stempel AG type foundry.

Out of his love and enjoyment of calligraphy, he produced in 1949 a magnificent book of 25 calligraphic alphabets, *Feder und Stichel* ("pen and stylus"). His natural feel for letterforms and his inspirational way of combining calligraphic vigor with typographic discipline, together with the wide range of his work, made him a master type designer. Herman Zapf has enjoyed a long and distinguished career in America and Germany as calligrapher, type designer, typographer, consultant, and teacher. In 1977, he became Professor of Typographic Computer Programing at Rochester Institute. Zapf has embraced the diverse technologies of metal, film and digital imaging, and made a major, enriching contribution to the world of graphic arts and typographic communication.

klmnopq

Among the many typefaces he designed are Michelangelo, Palatino, Sistina, Melior, Saphir (Sapphire), Aldus, Kompact, Optima, Zapf Dingbats, and Zapf Chancery.

Erik Spiekermann
One of the most active designers of the current typographic era, Erik Spiekermann, an internationally renowned German type designer and typographer, has for many years been fully immersed in all aspects of graphic design, typeface design, and typography. After starting a career in printing and typesetting, he studied art history, taught at the London School of Printing, and practiced freelance design. Through his graphic and typographic work, and as teacher, lecturer, and writer, Spiekermann has become a leading authority in type design. He was a founder of the major font supplier, FontShop International (FSI), and coedited FSI's famous yellow *FontBook*. FSI grew from Spiekermann's original concept: the company supplies fonts and typographic technical services, and also publishes the multimedia experimental typography magazine, *FUSE,* the brain-child of Neville Brody.

Spiekermann founded his design company, MetaDesign, in 1979. Germany's largest design group, it now has offices in Berlin, Zurich, San Francisco, and London. His book, *Stop stealing sheep, and find out how type works,* cowritten with E. M. Ginger, has become essential reading for anyone wanting to know the whys and wherefores of typographic design. His type designs include ITC Officina Sans, ITC Officina Serif, FF Meta, and Lo Type and Berliner Grotesk (adaptations).

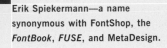

Erik Spiekermann—a name synonymous with FontShop, the *FontBook*, *FUSE*, and MetaDesign.

Neville Brody
A 1979 graduate of the London College of Printing, Neville Brody rapidly earned a reputation of being something of a rebel as well as a typographic star through his modernist and innovative typographic style. He challenged graphic conventions in his production of cutting-edge record covers and art direction of the British magazine, *The Face,* and London's *City Limits.* His antitraditionalist perspective put a unique and indelible stamp on

Neville Brody—one of the superstars of graphic design who have influenced much of our contemporary view of radical, innovative typography.

efghi

BLUR

efghi

TYSON

a design world bursting with energy and exploring the new medium of digitally generated design. His unconventional style was copied extensively by magazines other than his own, as well as an extraordinary range of consumer-associated graphics.

Brody also designed highly original typefaces. His designs include FF Blur, FF Auto-trace, FF Dirty faces, FF Gothic, and FF Pop. Behind his style and idiosyncratic letter-forms lies a formidable understanding of dynamic shape and form, which must identify him as one of the great graphic designers of his era. His work, commissioned by well-known companies and organizations across the world, has received international acclaim.

He is one of the founding partners in FontShop International (FSI), which markets a large and comprehensive range of fonts, and Brody, in particular, was the inspiration and pioneering energy behind the development of *FUSE*, FontShop's publication dealing with experimental typography. He continues to work very actively throughout the typographic and electronic communications design field.

Zuzana Licko

The impact that *Emigre* magazine had on the design world has already been discussed, but it is appropriate to mention Zuzana Licko again since, exceptionally, in her work as a type designer she consciously rejected the traditional calligraphic forms to which most typefaces owe their beginnings. Instead, using electronic media as her conceptual starting point, she began her design right where the pixel was—on the screen. For many who design for the new media, her work is truly inspirational and proves beyond a doubt that working within strict limitations can fuel, rather than curb, innovative energy. Recently, Licko has involved herself with the design of more conventional text typefaces (albeit with an inevitable "difference") to support the increased in-depth editorial content of *Emigre* magazine. Mrs. Eaves, her revival of Baskerville (named after John Baskerville's wife), shows evocative detailing particularly in the extensive set of ligatures, designed to stimulate visual interest.

Among Licko's typefaces are Oakland, Emigre, Emperor, Matrix, Modula Sans and Serif, Universal, Lunatix, Senator, Citizen, Elektrix, Totally Gothic, Totally Glyphic, Journal, Oblong and Variex (with Rudy VanderLans), and Triplex (with italic by John Downer).

Zuzana Licko, a post-computer designer, whose flair and understanding of low-resolution letterforms has produced many interesting and powerful typefaces.

abcdefghijklmn
opqrstuvwxyz

fffiflffiffl ~ ← →
¼ ½ ¾ ⅛ ⅜ ⅝ ⅞ ⅓ ⅔

MRS. EAVES LIGATURES AND FRACTIONS

Display Type

Display Type

The power and versatility of digital type software can be demonstrated at their greatest in working with display type. The possibilities of type manipulation are endless, limited only by the designer's imagination, creativity, and willingness to experiment.

While text setting is a composite of words, sentences, paragraphs, and columns, the sum of which creates its unique character, display type normally involves quite a small group of characters that form a single distinctive graphic element on the page. Within this element, the shape of each individual word or phrase takes on greater significance than in text setting, since it has a higher visual profile and impact.

Display type is generally considered to be any size above 14 point but, since sizes of type are relative to the context, the format in use, and the size of other graphic elements, this guideline should be interpreted loosely. Type that is not used for narrative text can be considered as display type; display type is used for short typographic announcements, statements, and signposting to attract attention.

Display elements may include some of the following:
- Punchlines and straplines in advertising material
- Titles, section names, chapter headings, and article or story titles in books
- Headlines, headings, subheadings, and cross-headings in magazines, newspapers, manuals and reports
- Web-page titles, main statements, punchlines in Web-page titles
- Signage systems and cartographic notations

The aim of display type in all forms of visual communication is that of attention-grabbing and highlighting, introducing key thoughts and ideas effectively, and directing the reader to absorb information in an efficient, structured manner. Display type can act as a powerful signposting element, bringing cohesion, variety, visual interest, and graphic impact to all forms of printed or screen-based pages. In working with display type, opportunities

A page from the magazine *U&lc* showing the use of the display type ITC Werkstatt. The face was inspired by the work of Rudolph Koch whose calligraphic typefaces were interesting in that the counter shapes bore no resemblance to the external shapes of the letters. Other unusual characters include the double rule hyphen, used here in different weights, styles and sizes.

ITC Werkstatt is a result of the combined talents of Alphabet Soup's Paul Crome and Satwinder Sehmi and ITC's Ilene Strizver and Colin Brignall. It is inspired by the work of Rudolph Koch, the renowned German calligrapher, punchcutter, and type designer of the first third of this century, without being based directly on any of Koch's typefaces. Werkstatt has obvious affinities with the heavy, woodcut look of Koch's popular Neuland, but also with display faces like Wallau and even the light, delicate Koch Antiqua. "Koch's unique typeface design style struck a chord with us," says Brignall, and he and Sehmi undertook an exhaustive study of Koch's typefaces and calligraphy before beginning their own design. Brignall began by drawing formal letters with a 55mm cap height, which Sehmi re-interpreted using a pen with a broad-edge nib. "Not an easy process," says Brignall, "since one of the features of Koch's style is that while it was calligraphic in spirit, most of the time his counter shapes did not bear any resemblance to the external shapes, as they would in normal calligraphy. This meant that Sehmi could not complete a whole character in one go, but had to create the outside and inside shapes separately and then ink in the center of the letters." The process was repeated, only without entirely filling in the outlines, for the Engraved version. Paul Crome handled the scanning and digitization, maintaining the hand-made feel while creating usable digital outlines. "The collaboration of artisans with particular skills," says Brignall, "in a modern-day, computer-aided studio environment, seems very much in step with the 'workshop' ethos that Rudolph Koch encouraged and promoted so much." www.itcfonts.com/itc/fonts/full/ITC258I.html www.itcfonts.com/itc/fonts/full/ITC258a.html **FONTEK.**

O%
interest, that's all we charge!

Pay **0%** on all purchases

Man

" A self-balancing, 28 jointed adapter-based biped; an electrochemical reduction plant, integral with segregated stowages of special energy extract in storage batteries for subsequent actuation of thousands of hydraulic and pneumatic pumps with motors attached; 62,000 miles of capillaries... The whole, extraordinary complex mechanism guided with exquisite precision from a turret in which are located telescopic and microscopic self-registering and recording range finders, a spectroscope, etc; the turret control being closely allied with an air-conditioning intake-and-exhaust, and a main fuel intake..."

The above definition was written by
R. Buckminster-Fuller in 1938

Above are examples of the use of each of the display elements discussed: size, weight, and disposition.

abound to incorporate fun, excitement, mood and character, color and decoration, as the context requires.

SIZE, WEIGHT, AND DISPOSITION

Contrast and good document signposting can be achieved easily through the application of fundamental typographic principles. Contrast of size, weight, and strategic positioning can all be used to good effect. The most sober of messages or publications can be lifted by the judicious use of well-ordered layout and thoughtful display type. It is possible to create powerful and easy-to-read information by the considered use of these methods of contrast. Not all the techniques should be used together. For example, if a heading has been set in a highly contrasting type size to create emphasis, it does not necessarily follow that it needs to be set in bold and in color as well.

Size

It is important to use enough contrast in size to make the required effect. Without sufficient contrast, a small change of size for headings may just look like a visual mistake. As an example of good contrast, if most of text matter in a publication is set in 8-pt type, a display size of 12 point would be appropriate.

Weight

The contrast in appearance must also be clearly evident when using different weights of type. Light type as well as bold can be used to lend emphasis. For example, a small amount of light type in a mass of bold can stand out just as well as bold type in a mass of lightness.

Changes in type weight alter the amount of ink on the page, which alters the mass of the printed elements, affecting layout balance. If space is at a premium, an advantage of using a strongly contrasting type weight is that it lets display lines sit close to the text without causing visual confusion.

Disposition

Quiet visual significance may be achieved by considered positioning or placement of text and display material. The contrast of a few words set in an open space or pulled out from the ranks of paragraphs can be powerful and effective. The positioning of any graphic element on a page is a central concern of the design process, but the often-forgotten fact should not be overlooked that placement alone can create tensions and contrasts that work extremely well. All the graphic elements on a page are related to

each other by either distance, mass, color, or a combination of all three. Visual tensions among elements depend on their relative proximity and shared points of alignment. To make display type work well, titles, heading, and subheadings should have a greater dynamic relationship to the individual parts of the information to which they refer. Dynamic grouping is the key to well-structured typeset information. Many of the methods used to achieve this are dealt with in paragraph formatting, where values for leading, indents, tabs, and other horizontal and vertical spaces can all be controlled.

DESIGNING, CONTROLLING, AND MODIFYING TYPE FOR DISPLAY

The very nature of display type demands that it should stand out. It should have a quality that on the one hand complements while also contrasts with the other elements on a page. The decision as to how much contrast and individuality that should be accorded to a piece of display type should be informed by the job in hand.

Here are just some of the modifications to type that digital software allows.

Display type may be:
• Extended or compressed
• Angled or sloped
• Rotated
• Colored—flat color or graduated tint
• Reversed-out from a color or black background
• Kerned
• Tracked loose or tracked tight
• Underlined or overlined
• Given drop shadows
• Outlined
• Framed
• Given tone, texture, or shading
• Pixelated
• Blurred
• Given three-dimensionality
• Given perspective
• Decorated or embellished
• Distorted
• Set in mixed typefaces
• Set in mixed case

Cover design by Eugenie Dodd Typographics.

Shown in the background are examples of the modifications to display type made available by digital software.

AUTO KERNING

RAILWAY

AUTO KERNING WITH +4 TRACKING (0.004 EM)

RAILWAY

AUTO KERNING WITH +4 TRACKING AND EXTRA PAIR KERNING

RAILWAY

+0   +0   +9   +2   +3   +3

AUTO KERNING

Diagonal

AUTO KERNING WITH −12 TRACKING (−0.012 EM)

Diagonal

AUTO KERNING WITH −12 TRACKING AND EXTRA PAIR KERNING

Diagonal

+7 +2 +0 −1 +1 +2 −2

The examples above show the effect of using various horizontal spacing controls. The top word is set with auto kerning default settings. By minus-tracking globally some of the letters are too tight. The bottom example shows overall minus-tracking with extra pair kerning.

Private view invitation by Eugenie Dodd Typographics.

Kerning and Tracking

Spacing between letters may be quite comfortable at reading or body-text sizes, but typically appears excessive when text type is set larger. Any discrepancy in spacing between individual character pairs will also be visually more obvious. Careful kerning and tracking must be considered to achieve good-looking, coherent word shapes. It is better to apply tracking first to achieve the right "color" from the overall shape before kerning individual letter pairs to harmonize the complete display. Larger character sizes can be tracked very tight, to the point of touching or overlapping, without losing legibility or impact. Too much loose tracking will begin to break up coherence and consequently affect readability, but it may let the letters play a more decorative role.

The tracking editors in QuarkXPress and InDesign may be used to adjust large type sizes for display work. A tracking value can easily be incorporated into a style sheet, which will be easier to handle in the long run.

Care and close attention to kerning at large sizes are essential. Poor kerning will be immediately detectable, and is the most frequent cause of unhappy-looking display work.

Type and Color

The color of type, in typographic terms, generally refers to the tonal quality that a mass of black printed text creates on a page. However, text can actually be colored to modify its visual impact greatly. Color control is one of the strongest tools that can be used to modify the behavior of graphic elements.

In display work, color can be used to help achieve many objectives. Words set in the same face, size, and weight may be colored to signify additional information. All the main graphics programs allow colors to be chosen using hue, saturation, and brightness values, which is invaluable when preparing a range of colors of equal tonal value. The same model may be also used to mix a palette of colors to create information structure. Through changes in color vibrancy, a hierarchical level of headings, otherwise similar in size and weight, can easily be created.

There is also a visual trade-off between mass and ink color. Large, heavyweight letters printed in pale colors recede, appearing lighter and less dominant. Color change can be used to alter strength, volume, and tone. Graduated color-tinting of type can provide a form of underscoring, whereas placing color behind type will greatly alter its significance.

Text in Drawing Applications

Adobe Illustrator and Macromedia Freehand provide many tools for the creation of custom display type. These programs are vector-based, which allows type outlines to be created and then manipulated. Filters and controls can be used extensively to generate unusual and exciting imagery, or

Some Pluses

Here are some of positive effects that can be achieved by different treatments of display type:

- Type set at an angle can break visual monotony.
- Greatly angled (over-italicized) type can produce a sense of urgency or speed.
- Type set round a circle can create a regular, well-defined graphic element.
- Blurring characters can soften the mood.
- Condensed or compressed type suggests a staccato effect and urgency.
- Expanded type can create a leisurely feeling.
- Textured type can give the page tactile quality.
- Outlining a letter gives it lighter mass.
- Colored type looks softer than black and appears to have less mass—it can be used at larger sizes.
- Drop shadows lift type from the surface of a page. Blurred drop shadows do it better.
- Black type printed over a very pale color reads better than black on white.

Some Minuses

More obvious methods of accentuating display type can produce negative results. For instance:

- Automatic underlining often strikes through the descenders of lowercase characters and distorts the word shape.
- Tints over key words often have the effect of making them less noticeable.
- All-capitals setting slows down readability and comprehension.
- Outlined type is hard to read.
- Applying complementary colors to different display type elements on a page sets up visual competition rather than contrast.
- Box rules around display type can interfere with readability and comprehension.
- Reversing type out of dark colors makes type look smaller.

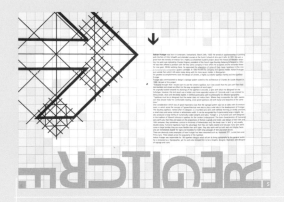

Expressive use of type demonstrating the qualities inherent in the letterforms. Examples by Anna Leaver, above, and Ed Jacob, right.

Film title by Marion Fink and Ben Ducket.

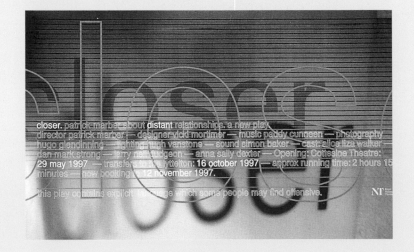

the manipulation may be subtle, and perhaps used to create individual ligatures or logotypes. Rotation, mirroring, inversion, distortion of type, placement of text along a path or around a circle, as well as perspective, outlining, drop-shadow effects, and the filling of letterforms with graduated tints, patterns, and even pictures are all possible.

Creating interesting display type in a drawing (vector-based) application has the distinct advantage of infinite scaling and maintained quality of output, whether you print from the software that created the effect or import it to another program such as QuarkXPress or Adobe InDesign.

Vector-based display type can easily be exported to and used by Web software, including Adobe GoLive, and Macromedia Flash, Dreamweaver, and Director.

TEXT IN PHOTOSHOP

Creating display type in Photoshop is easy and fun to do. Most of the coloring and distortions available in the drawing applications can be carried out in Photoshop, with the added benefit of airbrushing, an extensive range of filters and three-dimensional filter plug-ins.

The latest versions of Photoshop have many Web-related features that enable the quick conversion of display work to minimum-size files using dependable Web-safe colors.

EFFICIENCY OF DISPLAY TYPE

Customizing display type is fun, but the effect it may have on legibility and readability when positioned in a layout should be kept in mind. When you modify display type, reduced legibility and readability must be balanced against the power of contrast and graphic impact.

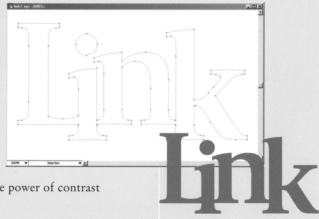

Overlapping letters fused into a single outline object using Pathfinder facility from a drawing program.

HOW MUCH DO WE NEED TO CUSTOMIZE OUR DISPLAY TYPE?

There are four fairly distinct levels to which we can control and customize display type.

First, we can employ all the characteristics of size, weight, and disposition of type to achieve strong, dynamic groupings of display type in relation to text.

Second, in conjunction with these techniques, we can tidy up visual inconsistencies through tracking and kerning.

Third, we can give display type individuality by customizing it with nondistorting techniques; for instance, using color, drop shadows, colored backgrounds, creating simple ligatures, or adding detail.

Fourth, we can make radical changes to display type by breaking up the letterforms and by using filters and software-generated distortion. The results tend much more toward decoration and mood-setting than towards readability.

We can use all four levels either independently or in carefully selected combinations appropriate to the context. Cultural style and fashion will influence display type but, no matter how the type is generated and utilized, the successful end result should be determined by the fundamental principles of typography—size, weight, and disposition. However creative or complex the required modification to display type may be, you can be sure that with digital wizardry there will be no limits.

Glossary
& Index

Glossary

Accent Mark added to a letter to indicate pronunciation, obtained by keying composite keystrokes (Macintosh) or by ALT + numeric code (Windows).

Adobe Font Metrics (AFM) *See* Metrics.

Adobe Multiple Master PostScript font format that allows two master outlines to be encoded as opposite ends of a design axis (e.g., weight, width, optical scale). The user chooses a point on the axis to generate an instance of the type with more or less of the axis feature. One axis requires two outlines, two axes require four, and three axes require eight. Adobe has ended development of the format but still sells the fonts it made.

Aliasing Jagged rendering of bitmaps owing to low resolution.

Aligned left Type arrangement in which all lines start at the same distance from the left of the page. Line ends are thus ragged. Also called flush left, ranged left.

Aligned right Type arrangement in which all lines end at the same distance from the right of the page. Line starts are thus ragged. Also called flush right, ranged right.

Ampersand The symbol "&," meaning "and."

Antialiasing Shading applied to pixels at the edges of characters to eliminate jaggedness and simulate smoothness.

Antiqua 1 Seriffed letterform. **2** Old Style letterform. **3** Roman type, rather than black letter.

Apex Junction of stems at the top of a letter, as in capital "A," "M," and "W."

Arabic numerals Numerals of Indian-Arab origin, i.e., 123, as opposed to Roman (CXXIII).

Arm Short horizontal, oblique, or curved stroke, as in capital "E," "L," "S," "K," and "Y."

Ascender Vertical stroke rising above the x-height in lowercase characters, as in "b," "h," "k," and "t."

Ascender height The height of the tallest ascenders in a type, above which there remains some body clearance.

ASCII A data or file format containing text and paragraph markers but no formatting information.

Asterisk Star-like symbol at ascender height, usually referring to a note.

Auto-leading Default line spacing in typesetting software, usually factory-set to 120% of the type size.

Autotrace Automated tracing function in some drawing software to convert a bitmap image to a vector outline.

Axis 1 Fixed reference line that allows position to be described. The x axis is horizontal, and the y vertical; both are measured from a point of origin, usually the top left of the page. **2** A continuum between two extremes, especially in a Multiple Master font, e.g., the bold–light axis.

Baseline An invisible line on which characters sit. All layout software calculates the position of type in relation to this line.

Baseline grid An invisible grid of horizontal lines at fixed, user-definable intervals, on which all type can be made to sit.

Baseline shift A software function applied to selected characters to move them above or below the baseline.

Bézier curve A curve or line between a starting point and an ending point, whose path is controlled by one or more further points (Bézier control points) according to a mathematical equation.

Bézier control point A point that determines the path of a line or curve connecting two points (i.e., a vector). The point can be adjusted on-screen, to reshape the Bézier curve by use of node-handles.

Bitmap font A font that describes character shapes in terms of which pixels in a grid should be switched on or off.

Black letter Class of heavy script-like traditional letterforms.

Body In metal type, the size of physical type, i.e., from the top of the ascenders to the bottom of the descenders, plus a little extra space (body clearance) above and below. In DTP, type size.

Body clearance The small amount of space above the ascenders and below the descenders that ensures that characters on consecutive lines of type do not touch.

Body copy Continuous text, excluding all forms of heading and displayed type.

Bold face Typeface design with heavy, wide strokes.

Bowl Fully closed round or oval stroke, e.g., in lowercase "b," "g," and "o."

Bracketed serif A serif joined to a stem by a smooth curve.

Bromide White photographic paper used in phototypesetting on which exposed characters appear black.

Browser Software that displays data and graphics according to instructions written in HTML. The best-known are Microsoft Internet Explorer and Netscape Navigator.

Bullet Solid circle, square, or dot placed in front of text to mark the start of a paragraph, list item, etc., and to highlight information.

Bureau Specialized prepress service that produces output from computer files at very high resolution and (if applicable) with accurate color. Output may be proofs, film for platemaking, or even plates.

Calligraphy The art of elegant writing and penmanship.

Camera-ready artwork Assembly of black-and-white images, including type, on a flat sheet or board ready to be photographed for platemaking.

Capitals Letterforms derived from Roman inscriptions. Often called caps or uppercase.

Cap height Height from the top of the capital "H" to the baseline. Capitals are often smaller than letters with ascenders.

Caps, small Capital letters of the same height as lowercase letters, and of height designed to harmonize with them. Mostly found in "expert" fonts. Small caps can be simulated by reducing normal caps, but they look lightweight and do not harmonize with the lowercase.

Cascading style sheets (CSS) Style sheets written in HTML and used on Web sites to give all pages a consistent appearance.

Casting off Calculation of how much space handwritten or typewritten copy will occupy when typeset. Almost obsolete.

Cathode ray tube (CRT) The display component of a computer monitor or television.

CD-ROM Compact Disc, Read-Only Memory—a common medium for distributing files, fonts, and software.

Centered A line of type arranged so that each end is equidistant from a central axis. Several centered lines in sequence will have ragged left and right edges that form a symmetrical shape.

Character Any type letterform, number, punctuation, or symbol.

Character assembly Typesetting, hand setting, or composition.

Character set The complete set of letters, numerals, punctuation, and symbols in a font.

Composition Typesetting, especially of body copy.

Composition sizes Text sizes suited to body copy, i.e., up to 14 point. *See also* Display type.

Compositor An operator who input text and had no design responsibilities. Now obsolete.

Color The tonal value created by a body of text on a page.

Condensed A narrower variant of typeface designed to ensure that weights of horizontal and vertical strokes stay balanced. User changes to horizontal scaling destroy this balance.

Contrast Planned differences in typographic size, color, weight, or disposition, used to show structure or add visual interest.

Copy Written material before it is typeset.

Copyright Legal right to control the use of intellectual property or created material, e.g., through licensing. Symbol: ©.

Core fonts Microsoft fonts issued widely with system and Internet software, designed to provide on-screen clarity and smoothness on any platform.

Counter The partly or fully enclosed area of a letter, as in lowercase "a" and "e," capital "B," etc.

Crossbar A horizontal stroke connecting two stems, as in capital "A" and "H," or a simple stroke, as in lowercase "f" and "t." Also called cross-stroke.

Dash Punctuation mark. The en dash (–) can mean "to" (e.g., Nice–Paris, 1950–55). The em dash (—) is most often used as a strident comma.

Dazzle The sparkling visual effect created when type with extreme stroke contrast is set in large text blocks.

Default Standard or factory settings in software that can usually be overridden by users.

Denominator In a vulgar (non-decimal) fraction, the bottom number, which expresses the parts of the whole, e.g., in ¼ the denominator is 4.

Descender The part of a lowercase letter projecting below the baseline, as in lowercase "g," "j," and "p."

Descender depth The lowest extent of the longest descenders in a type (below which there remains some body clearance).

Desktop publishing (DTP) Word-processing, typesetting, image manipulation, and page assembly done on a personal computer.

Diacritic An accent or mark above or below a character to indicate special pronunciation.

Dialog box A graphical interface in which a user enters instructions.

Didot point *See* Point.

Digital typesetting The imaging of type on screen and paper by use of bitmaps or outlines of characters stored electronically.

Dingbat A pi character.

Diphthong One glyph that represents two vowels, e.g., Æ.

Direct digital imaging Direct exposure of type and images onto a lithographic printing plate, with no intermediate film.

Discretionary hyphen Manually inserted hyphenation point to tell software where it must break a word if the whole word will not fit on a line. No hyphen will be visible unless the word is broken. Inserted at the start of a word, a discretionary hyphen prevents hyphenation.

Display Monitor or screen.

Display type Large type, usually above 14 point, used for headings, etc. *See also* Composition sizes.

Dry-transfer lettering High-quality self-adhesive plastic letters that can be transferred to artwork by rubbing them down, made by Letraset, *et al.*

Dots per inch (dpi) The resolution used to display an image. It depends on the display device's resolution. *See also* Resolution.

Drawing application Software that describes images and type in terms of scalable vectors.

Drop cap A large capital to start a paragraph. It hangs from the cap height of the first line.

Drop shadow A copy of a graphic or typographic element in a different tone, placed behind and offset from the original to suggest depth.

Ear The stroke attached to the bowl of lowercase "g." Also used for the angled or curved stroke of lowercase "r."

Egyptian Wide, slab-serif typeface.

Ellipsis Three points used to show a pause or a trailing off.

Em The square of the type size (body height), used as a relative unit of width, i.e., in 9-pt type it is 9 point, etc. A pica is 12 point, so a pica em is 12 points wide.

Em space 1 A word space the width of an em, i.e., in 9-pt type it is 9 point, etc. **2** In QuarkXPress, by default, the width of two Os (zeros): See *En.*

En A width unit equivalent to half an em. The width often given to lining numerals, so that they line up in tables.

En space A word space the width of an en, i.e., in 9-pt type it is 4.5 points, etc.

Encapsulated PostScript (EPS) A PostScript-language file format containing vector and (perhaps) bitmap information, with an encapsulated bitmap preview.

ESQ Enhanced Screen Quality: Monotype font technology for achieving optimum legibility on screen.

Exception dictionary A user-defined list of hyphenation exceptions. *See also* Hyphenation exceptions.

Expanded A wider variant of a typeface specially designed to ensure that horizontal and vertical strokes stay balanced. User changes to horizontal scaling destroy this balance. May also be called extended.

Expert set Supplementary font of characters extra to those in the related standard font, e.g., nonlining numerals (Old Style Figures, OSF), small capitals (SC), fractions, and symbols.

Eye The counter in "e."

Face Typeface, i.e., the visual appearance of a type design.

Family Related variants in a type-design concept. Common variations are by inclination (roman or italic), weight, and width. Thus Gill Sans Light, Gill Sans Medium, Gill Sans Bold, Gill Sans Light Italic, Gill Sans Medium Italic, and Gill Sans Bold Italic are family members.

Fat face Class of exceptionally bold type with exaggerated contrast of thicks and thins.

File formats Standards for encoding, storing, and exchanging data. Some formats have limitations, e.g., TIFF files contain no vector data, ASCII files lack formatting, and some graphics formats cannot have transparent backgrounds or support limited colorspaces.

Filmsetting Phototypesetting method that rendered type by shining light onto photographic paper through letters held in negative on a strip or disc.

Figures Numerals, numbers.

Fixed word spacing Identical space between all words (achievable only by setting text unjustified).

Folio Page, page number.

Font, fount A complete collection of letters, characters, punctuation, symbols, etc., and metrics for a given type. Some typefaces require several fonts. Used loosely, a typeface. Fount, the traditional British spelling, is pronounced "font."

Fraction Specially designed or constructed characters made from a denominator and numerator divided by fraction bar or solidus.

Fraction bar Oblique stroke more nearly horizontal than the solidus, used to construct fractions. Not available in all fonts or on all platforms.

Garalde One kind of Old Style letterform.

Glyph The shape of a character, accent, or symbol irrespective of its name.

H&J Hyphenation and Justification, often treated as one entity since they are interlinked.

Hairline The thinnest stroke of a letter, or the finest rule weight.

Hanging cap A drop cap.

Hanging indent An outdent or reverse indent, i.e., the first line of a paragraph starts to the left of the following lines.

Hanging punctuation Punctuation placed just outside the real column width so as not to disrupt a flush margin.

Hard hyphen A real hyphen keyed with the copy. Unlike a discretionary hyphen or one added by auto hyphenation, a hard hyphen is always visible.

Hexagraphic Printing with six inks to achieve a larger gamut of colors than can be had from four-color process.

Hinting Instructions contained in a font to determine how to correct an outline that does not exactly fit a pixel grid, e.g., when type appears on screen at low resolution or when it is printed at less than 600 dpi.

High-resolution printer An output device that renders type or images at 1200 dpi or better, so that the eye perceives the contours as smooth.

Hot metal The Linotype and Monotype processes, in which type was cast from molten metal as it was composed. Often used to mean letterpress.

Horizontal scaling Using DTP software to increase or decrease the width of type while the height remains the same.

House style A company or organization's typographic and layout rules, perhaps part of a wider corporate identity scheme.

HTML Hypertext mark-up language. A coding system for instructing Web browsers how to assemble and display type and images on Web pages.

Humanist One kind of Old Style letterform.

Hyphenation Breaking words in appropriate places in order to make them fit on a line of type.

Hyphenation exceptions Words chosen by the user to be exempt from automatic hyphenation. The user records any permitted hyphenation points (or none) in an exception dictionary.

Hyphenation Zone A user-defined area on the right of a column of unjustified lines. When widened, hyphenation decreases and the raggedness of the right margin increases.

Imagesetter A high-resolution device that outputs type and images from digital files on photographic paper or film.

Indent To move type inward from the left margin.

Inferior *See* Subscript.

Inkjet printer Printer that forms type and images by spraying fine particles of black or colored ink onto paper or film.

Instance A usable font created from a Multiple Master font.

Italic A sloping, script-inspired companion to a roman type. *See also* Sloped roman.

Jaggies Visible stepping of pixels at low resolution, e.g., on screen.

Justification Adjustment of the spaces between words and characters to make lines of equal length.

Kerning Adjustment of the space between two characters to obtain a satisfactory fit.

Kerning pair Adjacent characters that need kerning.

Kerning table A list of kerning pairs and the space to be added or subtracted. Values are set by the type designer, but can be modified by a skilled user.

Latin Characters used in Western (i.e., most European) languages.

Leader A row of dots to guide the eye across a blank area to further text, e.g., in tables.

Leading 1 Additional space inserted between lines of type. **2** This additional space plus the type size, i.e., the distance from one baseline to the next. Also called line feed.

Leg Same as *Tail* in capital "K," and "R."

Legibility The quality of being decipherable, recognizable, and distinguishable. *See also* Readability.

Letterpress Printing in which paper is applied to raised inked surfaces.

Letterspacing Adjustment of space between letters, mostly in display type, to achieve optical harmony.

Ligature Letters combined as a single character, e.g., "ff," "fi," and "ffl."

Light face A typeface with thinner strokes than normal.

Lineale San serif type classification. *See also* Sans serif.

Link The stroke connecting the top bowl and the loop in lowercase "g."

Lining numerals Numerals that are all the same height (usually the height of capital letters) and width (usually an en) so that they will align in tables. *See also* Nonlining numerals.

Loop The lower bowl of lowercase "g."

Lowercase Small letters (as opposed to capitals) derived from handwritten minuscules.

Matrix A mold for casting metal type.

Measure The width of a column of text, or the maximum possible width of a line of text.

Metrics Measurements held in a font file to control the spacing of type on screen and in print.

Minuscule A handwritten letter first used in the seventh century, and the precursor of lowercase.

Modern Class of typefaces with distinct contrast between thicks and thins (often with hairline serifs), and with vertical stress.

Monospacing Allocation of the same space to each character in a font regardless of its intrinsic characteristics. *See also* Proportional spacing.

Morphing Mutation of one shape or form into another.

Multimedia Combination of several media, e.g., print, animation, and sound.

Multiple Master *See also* Adobe Multiple Master.

Negative leading Leading less than the type size. It may be needed when types designed for composition sizes are set at display sizes.

Nonbreaking space A special word space used to keep a pair of words together: if both words will not fit on a line, both are taken over. Useful for linking quantities to units.

Nonlining numerals Numerals with ascenders and descenders, and of heights that harmonize with lowercase letters. Often of nonstandard width.

Oblique Equivalent to italic in some sans serif types, but achieved by sloping the roman letters. *See also* Sloped roman.

OCR A style of type first meant for optical character recognition or machine reading, but now a style in its own right.

Offset lithography Printing by the application of an inked roller to paper. The image on the printing roller is itself "offset" from another roller carrying a printing plate whose image area is treated to accept oil-based inks. Nonimage areas are kept damp to repel ink, since oil and water do not mix.

Old Style Class of seriffed types more or less based on pen-drawn letters, with angled stress and low or moderate contrast between thicks and thins. Sub-classes include (in historical sequence) Humanist and Garalde. The term includes modern revivals and imitations of the style.

Old Style Figures Nonlining numerals.

OpenType Cross-platform font format allowing TrueType and Type 1 fonts to be enclosed in one "wrapper" and offering large character sets.

Origination All preparation and assembly of visual components.

Orphan The first line of a paragraph starting at the foot of a column or the end of a page.

Outline font Font containing vector-based character outlines.

Output resolution Resolution supplied by equipment used for final imaging.

Overshoot Small optical correction that type designers apply to curved letters, so that

Prince text, headline decorative.

their tops exceed the cap height or x-height, and their bottoms protrude below the baseline.

Photocomposition Keyboard input with filmset output.

Photolithography Offset lithography from filmset origination.

Pi character Symbols and pictograms supplied in pi or dingbat fonts.

Pica em *See* Em.

Pixel Smallest unit in a bitmap, whether recorded input (e.g., a scan) or output (e.g., on screen).

Pixelation 1 Division into pixels for storage or display. **2** Display of a very small number of very large pixels, producing the effect of insufficient image resolution.

Planography Printing from a flat surface, as in lithography.

Point The basic unit of type measurement, abbreviated pt. Strictly, there are 72.27 Anglo-American points to an inch, but in most DTP software (and in HTML and CSS) this is rounded to 72 points. Since 12 points make a pica there are 6 picas to an inch in DTP, etc. However, continental Europe uses a different point: the Didot. The Didot point (ptD, p, or d) is 0.376 mm, slightly larger than the Anglo-American point (0.351 mm, or 0.353 mm in DTP), and there are 12 ptD in a cicero. In the Didot system, point size is called *corps*.

PostScript Adobe's patented page description language that enables vector-based outlines to be rasterized efficiently.

PostScript font A font that contains outline information as Bézier curves. *See also* Type 1 font.

Preferences Dialog box that lets a user change default settings.

Printer font An outline font.

Proportional spacing Allocation of space to each character in a font according to its intrinsic characteristics, i.e., the width of the glyph plus side bearings in proportion. Narrow characters, such as lowercase "i," thus occupy less space than wide ones, such as capital "W." *See also* Monospacing.

Punch A letter cut in metal for punching out a mold from which metal type could be cast. Punch cutters often refined type designs at different sizes.

Quotation marks Curled or tapered marks to set off speech, etc. "Smart quotes" in DTP

software allow automatic replacement of foot and inch marks with the proper glyphs.

Ranged left *See* Aligned left.

Rasterization Conversion of outlines into dots (bitmaps).

Raster image processor (RIP) Rasterizing software or hardware.

Readability The qualities in a legible type that make for sustained comfortable reading.

Rendering Presentation of digital data in comprehensible form, especially on screen.

Resolution Density of pixels or dots in a specified area, stated as number per linear unit, e.g., per inch. Input resolution is a property of the file, stated as pixels per inch (ppi). Output resolution is a property of the output device, stated as dots per inch (dpi) or (for image-setters) lines per inch (lpi).

Ring Accent used in Nordic languages, as in "å."

River Connecting white space running through consecutive lines of badly justified text.

Roman 1 Upright letters, as opposed to italics. **2** Medium weight, as opposed to bold.

Rule A straight line used as a typographic element to divide information or create stress, or for ornament.

Runaround Instruction to flow type around the rectangular, circular, or freeform shape of an image, another text block, etc.

Running text Body copy, usually of some length.

Sans serif Class of type with no serifs.

Screen font Low-resolution bitmap font for screen display.

Screen resolution For CRT monitors, 72 dpi for Macintosh, 96 dpi for PC; for LCD displays, up to 150 dpi and constantly improving. Since screen fonts are tailored to the output device, screen resolution may also be expressed as ppi.

Serif Short stroke at the end of stems, arms, and tails; styles include beaked, bracketed, hooked, slab, spur, and wedge.

Set solid Set with leading equal to the type size.

Side bearings Areas on either side of a glyph that ensure adequate separation from neighboring glyphs.

Sloped roman Poor substitute for italic made by slanting the roman letters. *See also* Oblique.

Small caps *See* Caps, small.

Smart quotes *See* Quotation marks.

Smoothing Technique to make small on-screen text crisper and easier to read. *See* Antialiasing.

Soft return Line break instruction within a paragraph.

Software Computer operating procedures and directions that control hardware.

Solidus Oblique stroke, also called a slash. May be used as a fraction bar if none is available.

Spine Main stroke of capital or lowercase "s," excluding the arms.

Spur Small spike at the base of the stem in some designs of capital "G."

Stem Vertical and full-length oblique strokes in a letter.

Stochastic Random use of tiny, identically sized dots to convert continuous tone for printing. More dots make images darker, fewer dots make them lighter. Stochastic screening avoids the moiré and rosetting that can occur in conventional screening.

Stress In types with thick-thin contrast, the direction of the thick strokes, i.e., the axis perpendicular to the thin strokes.

Stroke Individual linear element of a character.

Style Plain, bold, italic, etc., versions of a named typeface. *See also* Family.

Style sheet Sets of typographic attributes used to style different kinds of text. By applying the style sheet, all attributes can be styled at once.

Subheading Secondary heading to break up running text into more digestible components.

Subscript Reduced-size character set near the baseline, e.g., 2 in H_2O. Traditionally called an inferior character.

Suitcase On a Macintosh, the file containing a Type 1 screen font or all of a TrueType font.

Superior character Reduced-size character that hangs from the ascender height, used for notes, etc., e.g., content^a.

Superscript Reduced-size character with a user-defined offset above the x-height, used in technical work, e.g., (here with a subscript too) X_{tj}^k.

Symbol Pi character. *See* Pi font.

Swash Ornamental flourish added to some letterforms.

Swell Thickening in a curved stroke.

Tail Short downward stroke, as in capital "R," "K," and "Q." *See also* Leg.

Tabular figures Lining numerals.

Terminal The end of a stroke.

Thicks and thins Varying stroke widths that evolved from the directional movement of the pen in calligraphy.

Tight Very close, e.g., with greatly reduced word spacing or tracking.

Tracking Adjustment to normal character spacing applied equally to a large amount of text.

Transitional Class of seriffed types that have vertical stress, more thick-and-thin contrast than Old Styles but less abrupt contrast than Moderns, and some bracketing of serifs. Baskerville is a fair example.

TrueDoc Bitstream font format that allows large character sets.

TrueType Font format that contains outlines that Windows and Macintosh operating systems rasterize on the fly for the screen.

Type 1 font PostScript font with hinting.

Type 3 font PostScript font without hinting. Processor-intensive and obsolete.

Typeface The designed appearance of a character set.

Type family *See* Family.

Typesetting The process of assembling type to form running text together with any associated display type.

Type style *See* Style.

Typographer Someone engaged professionally in the design or use of type as a means of visual communication.

Typography The design activity of creating or using type and letterforms.

Uppercase Capital letters.

U/lc Normal combination of uppercase and lowercase letters.

Umlaut Accent: two dots above a vowel, e.g., "ü."

Unicode International standard for describing a character set.

Units per em Relative unit of measure used in typeface design. TrueType has 2048 units per em; PostScript has 1000.

Unjustified Type that aligns on one edge only, the opposite edge remaining ragged.

Vector Line between two points expressed as a mathematical formula.

Venetian Humanist: one kind of Old Style letterform.

Vertex Apex.

Vertical alignment Arrangement of type at the top, in the center, or at the foot of an area (e.g., text box, column, or page).

Vertical justification Arrangement of type so that it fills an area from top to bottom, achieved by increasing leading and paragraph spacing.

Vertical scaling Using DTP software to increase or decrease the height of type while the width remains the same.

Vulgar fractions Two-storey nondecimal fractions.

WebFont Maker Bitstream software to allow fonts to be served along with Web pages.

Weft Microsoft software to let fonts be served along with Web pages.

Weight The relative lightness or thickness of stroke in a type.

Widow 1 The last line of a paragraph falling at the top of a column or a page. It may look untidy, and may be taken for a heading. **2** A very short last line (one word or two very short words) ending a paragraph.

Word break *See* Hyphenation.

Word space White space between two words, defined in font metrics, varying from font to font, and user-adjustable. Its width is constant in unjustified text but varies in justified text.

WorldType Agfa Monotype font format that provides for large character sets.

WYSIWYG Acronym for What You See Is What You Get (pronounced "wizzywig").

x-height Height of lowercase "x," which most influences the tonal value of text set in a typeface.

Xerographic An electrostatic method of imaging used in copiers and laser printers.

Index